T0192550

NEXUS NETWORK JOURNAL

PATTERNS IN ARCHITECTURE

VOLUME 9, NUMBER 1
Spring 2007

KIM WILLIAMS BOOKS

Nexus Network Journal
Vol. 9
No. 1
Pp. 1-158
ISSN 1590-5896

CONTENTS

One of the most profound relationships between architecture and mathematics is related to patterns: the patterns of geometric decoration are perhaps the most obvious, but there are patterns in proportions, patterns in the generation of a series of numbers such as the Fibonacci series, patterns in the construction of geometric lengths, and patterns in modular constructions as well.

This issue of the *Nexus Network Journal* is dedicated to various kinds of patterns in architecture. Buthayna Eilouti and Amer Al-Jokhadar address patterns in shape grammars in the ground plans of Mamluk *madrasas*, religious schools. The Mamluk sultans were a series of rulers who reigned in Eygpt for nearly 300 years, from 1250 to 1517, and whose reign saw the creation of very beautiful art and architecture. In their two papers in this issue, "A Generative System for Mamluk Madrasa Form-Making" and "A Computer-Augmented Precedent-Based Mamluk Madrasa Plan Generator", Eilouti and Al-Jokhadar first compare the significant forms of the ground plans of Mamluk madrasas (by studying sixteen madrasas built in Egypt, Syria, and Palestine during the Mamluk period) to create a shape vocabulary, then formulate the operational rules that govern combinations of the forms (a shape grammar), and finally, create an interactive computerized plan generator. The shape grammar permits realization of a myriad of patterns based on the initial vocabulary that all lie within the framework of the madrasa program requirements.

Giulio Magli goes back further in history, to the age of Greek colonies in Italy before they were conquered by the Romans, to examine patterns in urban design. While most Greek cities were built in a rectangular pattern, as were the Etruscan and the Roman, constructed around a set of orthogonal axes (the *decumanus* and *cardus*), the settlements examined by Magli in "*Non*-Orthogonal Features in the Planning of Four Ancient Towns in Central Italy" exhibit radial patterns. Magli links this kind of radial urban planning to the radial patterns of the cosmos.

In "Traditional Patterns in Pyrgi of Chios: Mathematics and Community," Charoula Stathopoulou examines the geometric patterns that decorate the buildings of the town of Pyrgi, on the Greek island of Chios, and uses the research methodology of anthropologists to examine the relationships between pattern and community there.

"Curve Fitting" is a study of ways to construct a function so that its graph most closely approximates the pattern given by a set of points. Dirk Huylebrouck's paper, "Curve Fitting in Architecture" examines how a pattern of points extracted from an arch, or the pattern of points that define the curve of a nuclear plant, might be associated to a precise mathematical curve. Pattern definition in this case could help resolve the issue of whether or not the architect intended to describe a precise mathematical curve in his or her construction. For instance, are the arched gates and windows in Gaudi's *Paelle Guell* in the shape of hyperbolic cosines or parabolas? Comparison of the pattern formed by a set of measured points on the architectural forms with the graphs of the functions could provide an answer.

Patterns in the architecture of Frank Lloyd Wright have been studied for some time, both as plan generators, as in the Palmer House, and as decorative motifs, as in his window designs. Piet Mondrian's paintings are well-known examples of patterns in art. In "Integrated Function Systems and Organic Architecture from Wright to Mondrian", James

1590-5896/07/010005-2 DOI 10.1007/s00004-006-0026-6
© Kim Williams Books, Turin

Harris looks at the designs of these two masters to extract the rules of their pattern generation and propose possible applications.

In the Geometer's Angle column, Rachel Fletcher examines geometric constructions that square the circle. Her "Squaring the Circle: Marriage of Heaven and Earth" looks at the combination of square and circular patterns.

This issue is completed by B. Lynn Bodner's report of the annual Bridges conference, the 2006 edition of which took place in London in August of this year, Sylvie Duvernoy's report on the symposium "Guarino Guarini's Chapel of the Holy Shroud in Turin: Open Questions, Possible Solutions", and by Kay Bea Jones's review of the exhibit "Zero Gravity. Franco Albini. Costruire le Modernità" in Milan.

I think that this is one of the best issues ever of the *Nexus Network Journal* and I hope you enjoy it.

Kim Williams

Buthayna H. Eilouti

Department of Architectural
Engineering
Jordan University of Science and
Technology
POB 3030
Irbid 22110, JORDAN
buthayna@umich.edu

Amer M. Al-Jokhadar

College of Architecture & Design
German-Jordanian University
Amman, JORDAN

TURATH: Heritage Conservation
Management and Environmental Design
Consultants
P.O.Box: 402, Amman, 11118 JORDAN
amerjokh@hotmail.com
amerjokh@gmail.com

Keywords. Shape grammar, generative
system, visual studies, Mamluk
architecture, school design, Islamic
architecture, design process

Research

A Generative System for Mamluk Madrasa Form-Making

Abstract. In this paper, a parametric shape grammar for the derivation of the floor plans of educational buildings (*madrasas*) in Mamluk architecture is presented. The grammar is constructed using a corpus of sixteen Mamluk madrasas that were built in Egypt, Syria, and Palestine during the Mamluk period. Based on an epistemological premise of structuralism, the morphology of Mamluk madrasas is analyzed to deduce commonalities of the formal and compositional aspects among them. The set of underlying common lexical and syntactic elements that are shared by the study cases is listed. The shape rule schemata to derive Mamluk madrasa floor plans are formulated. The sets of lexical elements and syntactic rules are systematized to form a linguistic framework. The theoretical framework for the formal language of Mamluk architecture is structured to establish a basis for a computerized model for the automatic derivation of Mamluk madrasa floor plans.

1 Introduction

On the one hand, the study of design, its underlying representations, and the methods that can be used to derive new artifacts are important research topics in many disciplines, including engineering and architecture. Form-making entails design activities that have a direct influence on the appearance of the artifacts produced. Its study involves the establishment of explicit and systematic links between the form of an artifact, its visual properties, its compositional attributes and its generative considerations. In addition, form-making is concerned with the processes and considerations that precede and follow as well as those that produce the final form. Shape grammar represents a systematic method for studying the form-making layer of design activities. Studies in the area of shape grammars are well established. In many applications they proved to be powerful in shape derivation, analysis and prediction. However, many of their potentials are still far from being fully explored, especially in the area of understanding the morphology of architectural precedents.

On the other hand, Mamluk architecture represents a significant period of Islamic architecture. It displays most of the aesthetic principles that underlie Islamic architecture. Most of the aesthetic values of Mamluk architecture are exhibited in Mamluk educational buildings (madrasas). The morphological structure that underlies the forms of Mamluk madrasas can be mathematically analyzed and syntactically systematized to formulate a powerful compositional language that may help in the understanding of the Mamluki style and the aesthetic principles of Islamic architecture.

There is no systematic study of the formal aspects of Mamluk architecture that explicitly articulates the compositional language that underlies its design, and the procedural sequence of its form generation.

In this paper, the two strands of shape grammar implementation and Mamluk madrasa morphological investigation are studied and interlaced. The connection between the two strands produces a new language for the formal composition of Mamluk madrasas. This language summarizes a morphological analysis of Mamluk educational buildings in a concise generative system that can be used to derive emergent examples of this significant period in the history of Islamic architecture.

2 Background

2.1 Formal Languages

Shape grammar is that part of design study which deals with the morphology of the overall forms of products and their internal structures and with the incremental processes that generate them. It is concerned with the constituent components of a form and their arrangements and relationships. It usually emphasizes the *lexical level* (vocabulary elements) and the *syntactical level* (grammars and relationships) of the architectural composition, rather than the *semantic, symbolic* or the *semiotic* levels. When rules of a shape grammar are executed on a given subset of vocabulary elements, different alternatives of artifact forms are created. The set of all forms that are generated by applying given rules on given vocabulary elements constitutes a formal language, which may correspond to a visual style.

Studies in the area of shape grammar started in the early 1970s, but their roots in pattern language, typology and systematic design methods started earlier. Although shape grammars have good predictive, generative, derivative, and descriptive powers, they do not focus on the historical, social, functional, or symbolic aspects of the architectural compositions.

Using a step-by-step process to generate a language of design, a shape grammarian typically formulates a set of shape rules, that is, transformations and parameters that can be applied to a set of given vocabulary shapes in order to reproduce existing shapes and come up with emergent ones. Different types of shape grammars can be defined according to the restrictions that are used in their applications. They vary according to the format of rules, variable parameters, constant proportions, allowed augmented attributes, and the order of rule applications. All components of shape grammars (vocabularies, spatial relationships, parameters, attributes, rules, transformations, and initial shapes) provide a foundation for a science of form-making and for a theory of systematic architectural design and composition methodology through algorithms that perform arithmetic calculations on geometric shapes.

Stiny [1980] proposed a framework to define a language of design, constructed by means of shape grammars. The framework can be developed through applying the following five stages [Osman 1998]:

1. **Vocabulary definition**. The basic shapes of a formal language are defined. They function as the basic building blocks for design.
2. **Spatial relationship determination**. In this stage, the structure observed in a set of designs is investigated to deduce the spatial relationships that are used to connect the building blocks.

3. **Rule formulation.** Shape rules are formulated in terms of the spatial relationships identified in the second stage on the basic shapes listed in the first stage.
4. **Shape combination.** Shapes in the vocabulary set are combined to form initial and subsequent shapes. Shape rules are applied recursively to initial shapes to generate new shapes.
5. **Shape grammar articulation.** Grammars are specified in terms of shape rules, initial shapes and new shapes. Each shape grammar defines a language of design.

2.2 Mamluk madrasas

Visually speaking, Mamluk architecture represents a significant period of the Islamic architecture. It exhibits the major goal of Islamic architecture, according to which designers strive to achieve harmony between people, their environment and their creator. In general, there are no strict rules defined to govern Islamic architectural design. Designers of the major institutional buildings of Islamic cities used local materials and construction methods and applied abstract geometric languages to express in their own ways the character, order, integrity, harmony, and unity of architecture. However, when the great monuments and precedents of Islamic architecture are examined, their formal structure reveals a complex system of geometrical relationships, a well-designed hierarchy of space organizations and a highly sophisticated articulation of ornaments, as well as deep symbolic and semantic connotations. Geometry in Islamic architecture was developed into a sacred science. It has been structured to express the Islamic beliefs and views of the relationships between the world, man, and God [Himmo 1995]. In the Islamic perspective, the method of deriving all the organizational proportions of a building form from the harmonious recursive division of a basic shape is a symbolic way of expressing the oneness of God and his presence everywhere [Himmo 1995]. Compositions in the Islamic architecture have been transformed into highly abstracted shapes on which principles of rhythmic repetitions, unity, symmetry, and variation in scale were applied to create ordered yet dynamic effects. Shape in Islamic architecture is strongly related to the study of mathematics and other sciences.

In Islamic culture, education has always been closely connected to worship. Expressing this, from their beginnings Islamic mosques have been used for both praying and learning. Over time, the mosque experienced various transformations of functions, which resulted in the establishment of a number of related building types of social, educational and religious characters and with narrower functional scopes. The two interrelated functions (worship and teaching) of mosques eventually diverged. The separation resulted in a distinguished sacred mosque and a madrasa. The early madrasa buildings offered special open and closed teaching halls. The form and function of the early madrasas were similar to those of the mosques. Eventually, the architectural typology of the madrasa forms became so prominent that it visually influenced mosque architecture in the Islamic region during the twelfth to the fifteenth centuries A.D. [Bianca 2000]. The two main prototypes of madrasa layout shapes are the open courtyard madrasa and the closed or domed courtyard madrasa (fig. 1). The domed madrasas are usually smaller buildings whilst those with an open courtyard are generally larger and have central Iwans surrounded by arcades.

The most common formal prototype of the early madrasas is the four-Iwan plan. An example of this prototype can be seen in the Mustansriyya Madrasa in Baghdad. Although it is traditionally thought that madrasas provided sleeping and working accommodation for students, the extant examples show that this was not a rule and it is only later on the progress line of madrasa buildings that student facilities became accepted parts of a madrasa design [Bianca 2000].

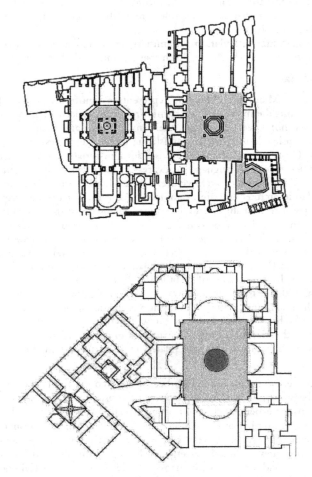

Fig. 1. The main madrasa layout types: a, above) The open courtyard madrasa type (al-Sultan Qalawun Madrasa in Cairo); b, below) The roofed or domed courtyard madrasa *dorqa'a* type (al-Sultan Inal Madrasa in Cairo). Reproduced by authors, from [Organization of Islamic Capitals and Cities 1990]

3 The morphology of Mamluk madrasa designs

3.1 Case studies of Mamluk madrasas

In order to develop a shape grammar for Mamluk madrasa design, a sample of sixteen cases has been assembled (Table 1). The sample consists of thirteen madrasas in Cairo, two

madrasas in Jerusalem, and one madrasa in Aleppo. The madrasas in the sample have been
selected for their important role as representatives of Mamluk architecture.

Type	No.	Case Study	Location
Type Two: covered court "dorqa'a"	1	al-Ashraf Barsbay Madrasa	Cairo
	2	al-Kadi Zein al-Dien Yehya Madrasa	Cairo
	3	al-Amir Kerkamas Madrasa	Cairo
	4	al-Sultan Inal Madrasa	Cairo
	5	al-Ghouri Madrasa	Cairo
	6	Umm al-Sultan Sha'ban Madrasa	Cairo
	7	Kani Bay Kara al-Remah Madrasa	Cairo
	8	al-Sultan Kaytebay Madrasa	Cairo
	9	Abu-Baker Mezher Madrasa	Cairo
	14	Tashtamar Madrasa	Jerusalem
Type One: open court	10	al-Thahei Barquq Madrasa	Cairo
	11	al-Sultan Hasan Madrasa	Cairo
	12	Serghtemsh Madrasa	Cairo
	13	al-Sultan Qalawun Madrasa	Cairo
	15	al-Saffaheyya Madrasa	Aleppo
	16	al-Baladiyya Madrasa	Jerusalem

Table 1: The study sample for Mamluk madrasas classified according to their geographic locations

3.2 Common compositional features of Mamluk madrasas

Most of the Mamluk madrasas in the sample were erected on restricted sites. The
exterior layouts of these madrasas respected the shape of the site they were constructed on.
Thus, irregular ground floor plan shapes were almost always generated. However,
considerable thought and effort were often given by the designer to make the building
regular in shape inside. Basic shapes were used as the basis for generating all interior spaces.
Most of the major interior spaces were oriented toward the "Qibla direction" (the prayer or
Mecca direction). Intermediate spaces appeared between the perfectly regular shapes of the
interior spaces and the irregular outer boundary of the site.

Typically, the aforementioned four-Iwan madrasa was a dominant prototype. In this
prototype, the four Iwans (South-Eastern Iwan or "Qibla Iwan", North-Western Iwan,
South-Western Iwan, and North-Eastern Iwan) surround the central courtyard. The other
spaces were located on the sides. In addition to each Iwan, facilities for many functions
have been designed: a residential unit for the sheikh (teacher), small units for students,
small court, *sabil* (free water fountain), *minaret* (tower), the tomb for the patron of the
madrasa, corridors and transitional spaces, *sadla* (secondary Iwan), ablution space, and
water closets.

Analyzing the sixteen case studies reveals the following major distinct elements:

- The dominant portal space.
- Great Iwans with vaulted roofing.

- Huge interior courtyards or *Sahns*.
- Sheikh's and students' cells.
- Large spaces covered with stone vaulting.
- Minarets that emphasize the portals.

3.3 Mamluk madrasa typology

In general, the floor plan of Mamluk madrasa exhibits a logical design with a vigorous articulation of clearly distinct elements. The design components include great teaching Iwans with tierce-point stone vaulting, huge interior courtyards, mausoleums under high cupolas, residential cells and polygonal minarets [Stierlin 1984]. The overall plan is shaped as a rectangular courtyard with an Iwan in the center of each side. Teaching takes place in the Iwans, and the students live in the cells arranged along the intermediate walls. The dominant architectural feature of this typology is the four Iwans built into the center of each courtyard side.

Later Mamluk madrasas tend to rise vertically in a number of stories, as opposed to the flat horizontal expansion of the early ones, which provided space for the multitudes on the same ground level. In addition, they tend to attach subsidiary units such as tomb chambers and *sabil-kuttab* to the main madrasa functions, converting their simple forms into large funerary complexes [Parker 1985].

3.4 Bahri and Burgi Mamluk madrasas

Mamluk sovereignty can be broadly classified into two periods: the *Bahri*, which ruled from 1250 A.D. to 1382 A.D., and the *Burgi*, which ruled from 1382 A.D. to 1512 A.D. [Parker 1985]. Mamluks spread extensively in Cairo, but they also reached Jerusalem and Aleppo.

Madrasas in the Bahri Mamluk Period are characterized by the following features (fig. 2a):

- The four Iwans surround the open courtyard, while other spaces are located on the sides and mainly in the first floor. This type has been used for teaching one or more legal rites;
- In addition to each Iwan, many facilities have been designed. These include a residential unit for the Sheikh (teacher), small units for students and small courts;
- In most madrasas, there was a tomb for the patron of the madrasa and his family;
- The portal space has been marked and emphasized by a minaret above it;
- The ablution and water closets are located in the back of madrasa for ventilation, and oriented to face the sun. Their level is lower than that of the madrasa itself.

Madrasas in the Burgi Mamluk period are characterized by the following features (fig. 2b):

- The madrasa consists of large and central open courtyard surrounded by four Iwans. These Iwans were divided into *riwaqs* (roofed halls), or covered by either pointed vaults or wooden roofs;

- The ground floor plan includes spaces for the family of the patron of the madrasa, and rooms for teachers and students;
- Another type in this period is the roofed courtyard *dorqa'a* which has been covered by a *Shokhsheikha* (wind ventilator) instead of the open court;
- The two lateral Iwans have been replaced by smaller Iwans;
- Most madrasas have a "*sabil*" (free water fountain);
- In many examples, the whole madrasa was used to teach one legal rite;
- The portal and vestibule space have been also marked and emphasized by a minaret above the main gateway;
- The ablution and water closets have been located in the back of madrasa.

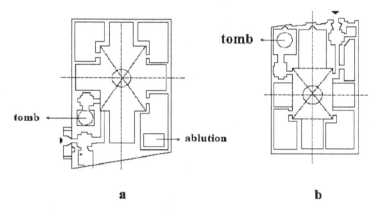

a   b

Fig. 2. Layouts of madrasa floor plans: a) in the Bahri Mamluk Period, b) in the Burgi Mamluk Period. Reproduced by authors, from [Organization of Islamic Capitals and Cities 1990]

Mamluk architects repeatedly used certain features in their buildings that shared some common attributes in size and type. Thus, close relationships can always be found among these buildings in their proportions, geometric shapes, topological relationships of spaces and formal compositional aspects. Some of the common features that can be observed in the Mamluk madrasas are:

- The geometric features and proportions of the great and small arches of building's *Sahns* (central courts of madrasa) are similar;
- The position of Iwans may be identified by the fenestrations of the façades of the madrasa without actually entering the building;
- Externally, fenestrations, whether alone or in groups, were set back from the wall planes in recesses. The dimensions and shapes of the openings and recesses reflect the position and size of the hidden Iwan and provide the details necessary to identify the Qibla (Mecca orientation) Iwan;
- The massing configurations illustrate apparent recourse to asymmetrical compositions.
- The façade treatment exhibits skillful articulation of openings and detailing of ornaments;

- The case study sample displays well-defined vocabulary components such as gateways, courtyards, Iwans, domed funerary hall and minarets;
- Locations of the main components such as court(s) and Iwan(s) are similar;
- A proportional system of individual components and their combinations underlies the case studies;
- The overall organizational rules that govern combining the spaces are based on the principles of symmetry, axis position and rotation, and the topological relationships of spaces.

4 A generative system for the Mamluk madrasa floor plans

A morphological analysis of Mamluk madrasa architectural examples reveals the existence of underlying coherent geometric language on all scales. Within this language, architectural vocabulary elements along with spatial organization rules and aesthetic principles of the components and their compositions are defined.

The resultant formal language can be represented as a parametric shape grammar that is based on a set of points, lines, and labels. Multiple aspects of the formal language can be parameterized. The parametric attributes include the size and angle of the rooms, the position at which courtyard walls are attached to rooms, the size of the courtyard and the arrangement and size of the elements of the building. Such a parameterization of rules significantly decreases the number of the proposed rules. It also increases the predictive and derivative powers of the formulated grammar.

4.1 The vocabulary elements of Mamluk madrasas

In making any architectural statement, the designer calls upon a formal vocabulary drawn from his or her previous experience and from the background tradition or culture in which the design is being executed [Serageldin 1988]. Among the institutions *(khanqah, ribat, and madrasa)* that were common in Mamluk architecture the one that describes most of its architectural vocabulary is the madrasa. The spatial vocabulary elements shared by the case studies are illustrated in fig. 3.

4.2 The Grammar of Mamluk madrasas

4.2.1 The procedure of grammatical rule formulation. The grammar rules are formulated according to the following considerations:

- Rules are formulated to specify how sub-shapes of a composition in progress will be replaced by other shapes.
- A rule applies if there is a similarity transformation that will bring the shape on the left-hand side of a rule into coincidence with a sub-shape in the shape vocabulary list.
- Labels are alphabetic characters that are associated with points and shapes. These labels are used to control the application of rules. An example of these labels is the symbol associated with the qibla wall which controls the Qibla direction of madrasa spaces.
- Shapes have proportional parameters which will be assigned by the grammar description during the process of its application.

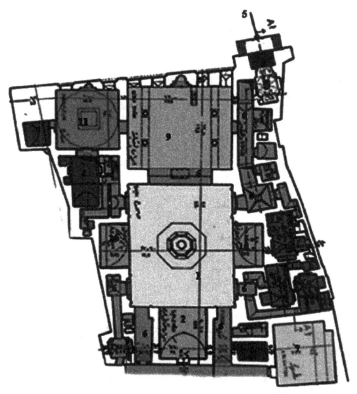

Fig. 3. The Vocabulary Elements of the Plan of al-Thaher Barquq madrasa, Cairo: 1– Courtyard (or *dorqa'a*); 2–North-Western *Iwan* (opposite to Qibla Iwan); 3–Derka "entrance spaces"; 4–Small Cells for the Students; 5–Main Entrance and portal space; 6–Corridors and Transitional Spaces; 7–Ablution Space; 8–Sabil (free water fountain); 9–South-Eastern Iwan (Qibla Iwan); 10–Southwestern and Northeastern *Iwans*; 11–Tomb or Mausoleum for the madrasa's patron; 12–Minaret (see fig. 4); 13–Teacher's House (Sheikh's House); 14–Northern and Southern secondary *Iwans (Sadla)*. (Reproduced by authors, from [Organization of Islamic Capitals and Cities 1990]

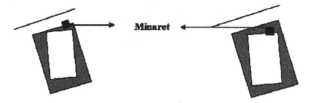

Fig. 4. A minaret that is: a) aligned with the street, while the rest of the building follows the angle of Mecca orientation, b) aligned with Mecca orientation

The morphological aspects, which include visual principles such as symmetry, proportion, geometry, axes distribution, addition and subtraction of spatial organizations, and shape transformations, are considered in the shape grammar development. These aspects are illustrated in fig. 5.

In general, the grammar derives the plans starting from organizing the exterior layout shape and the interior spaces, and proceeding down to the details of walls, doors, and windows. The grammar is formulated in a way that encourages a designer to start with the determination of the exterior layout. This point of departure is more efficient for controlling the overall shape, because the interior layout is more constant in its geometric shape and orientation. The overall structure of the formulated grammar is illustrated in fig. 6. The computational model is out of the scope of this paper. Its study is presented in a separate paper [Eilouti and Al-Jakhadar 2007].

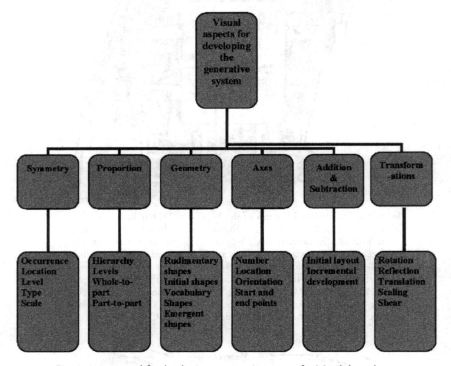

Fig. 5. Aspects used for developing a generative system for Mamluk madrasas

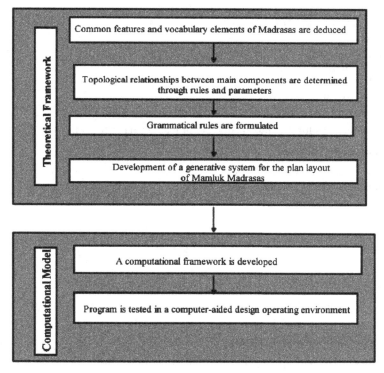

Fig. 6. A flowchart illustrating the different stages for developing a generative system for Mamluk madrasa floor plans

4.2.2. The grammar rules for generating Mamluk madrasa floor plans. All rules of Mamluk madrasas are numbered using the system $X - n - Y$, where:

X: is the name of the stage. The possible values of this variable are:
SL (Schematic Layout)
AC (Architectural Components)
WT (Walls Thicknesses)
OP (Openings)
TR (Termination)

n: is the rule's number in stage X.

Y: describes the sequential step that is associated with the rule (n). The possible values of this variable are:
PSH4 (the addition of a Parametric Shape with Four Sides)
or **PSH5** (the addition of a Parametric Shape with Five Sides)
or **PSH6** (the addition of a Parametric Shape with Six Sides)
or **PSH7** (the addition of a Parametric Shape with Seven Sides)
or **PSH8** (the addition of a Parametric Shape with Eight Sides)
or **PSRF** (the addition of a Parametric Shape and then Reflect it)

or **A** (the addition of a Parametric Shape - Arc)

or **C** (the addition of a Parametric Shape - Circle)

or **M** (Moving)

or **R** (Rotation)

or **D** (Division)

or **S** (Scaling)

or **U** (Union)

or **L** (Drawing Lines)

There are three steps for Mamluk madrasa floor plan form generation. These are:

1. Initial Shape Location:

The initial shape, from which all plans are generated, establishes a point labeled "A" at the origin of the coordinate system, as shown in fig. 7.

Fig. 7. The initial shape from which all Mamluk madrasa floor plans are generated: a labeled axis.

2. Exterior Layout Shape Determination:

The exterior layout shape is generated around the initial shape. Six alternatives for this set are defined. These are illustrated in fig. 8.

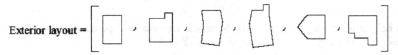

Fig. 8. The Exterior Layout Shapes

3. Grammar Rule Application:

The formulated grammar consists of 93 rules distributed as follows:

Rule type	Number of Rules
Rules for generating schematic external layout shapes	12
Rules for generating architectural components:	
Rules for generating interior spaces	2
Rules for determining the orientation of spaces	1
Rules for generating Iwans and courtyards	3
Rules for scaling Iwans and courtyards	3
Rules for generating spaces between Iwans	2
Rules for generating lateral spaces	4
Rules for generating mihrab in the Qibla-Iwan	3
Rules for articulating interior spaces	4

5 The derivation process of Mamluk madrasa floor plan

In order to demonstrate the shape grammars developed so far, an example is illustrated. It is selected from Type 2 (see fig. 1b), which has a roofed courtyard *dorqa'a*. This case is al-Ashraf Barsbay Madrasa in Cairo (1425 A.D.). The main components of this building are the central open courtyard surrounded by two large Iwans. The Qibla-Iwan is the largest one. There are also two other small Iwans, *sadlas*. The total area of this madrasa is approximately 1550 m², and the area of the central courtyard is about 230 m². Thus, the proportion between the total area and the area of the courtyard is 1:6.7. Other spaces that are represented in this madrasa are the mausoleum in the Eastern-North side, *sabil-kuttab*, and a minaret above the main portal.

The main entrance of this example is characterized by the vestibule space. The main gateway of the madrasa does not allow immediate access to the indoor spaces, but leads into a passage with a 90° turn, so that it is impossible to see the courtyard from the outside. This indirect access enhances the creation of calm learning environment inside. Another compositional feature is the division between the ablution space and other rooms. The main reason for this is the orientation to the wind, ventilation and sun.

The derivation process of this example is organized into five major stages. These consist of the generation of the exterior layout, the generation of architectural components, the determination of walls thicknesses, the assignment of the openings – doors and windows–, and, finally, the rule termination stage. Within these stages ten more detailed stages can be identified. Figs. 9-12 illustrate the ten stages through which the overall ground floor plan of al-Ashraf Barsbay Madrasa was derived. The ten stages are:

1. Determining the shape of the Exterior Layout by applying the rules: SL-1-PSH4, SL-2-A and SL-13-M (fig. 9a):

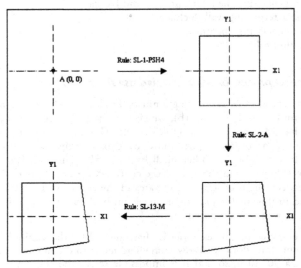

Fig. 9a

2. Outlining the Overall Interior Spaces by applying the rules: AC-3-R, AC-2-M, AC-1-PSH4 (fig. 9b):

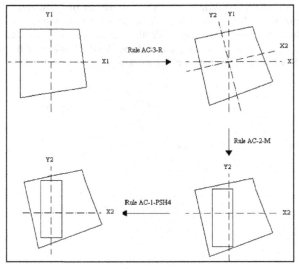

Fig. 9b

3. Generating the overall interior spaces by applying the rules from (AC-4-D) to (AC-9-S) (fig. 9c):

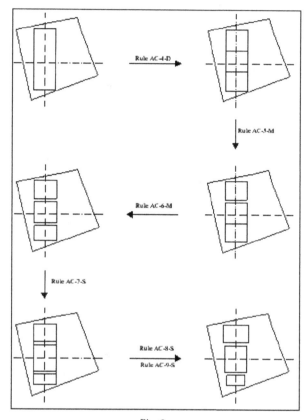

Fig. 9c

4. Generating the courtyards, Iwans, and minbar by applying the rules from AC-10-PSRF to AC-21-PSH4 (fig. 10a):

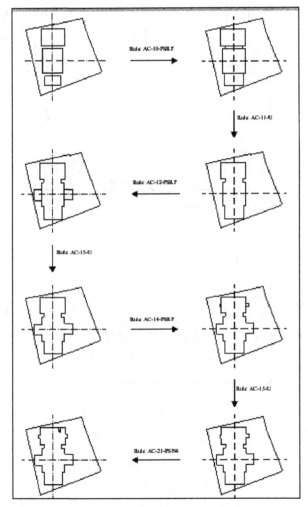

Fig. 10a

5. Generating *mihrab*, *sadla*, and tomb by applying the rules from AC-16-A to AC-24/b-C (fig. 10b):

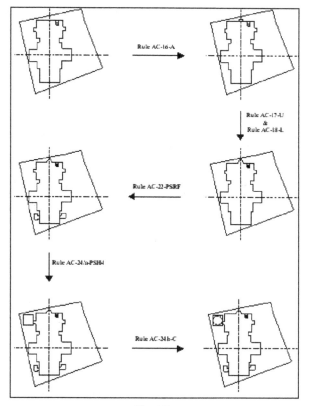

Fig. 10b

6. Generating ablution space, and cells around the courtyard by applying the rules from AC-27-PSH4 to AC-32-U (fig. 11a):

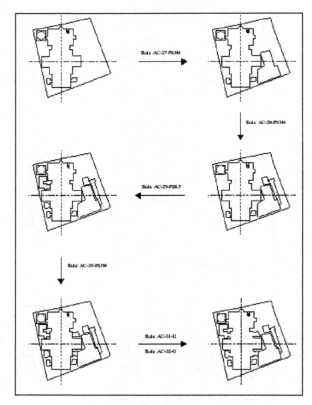

Fig. 11a

7. Generating the entrance and sheikh's house by applying the rules from AC-33/a-PSH4 to AC-40-U (fig. 11b):

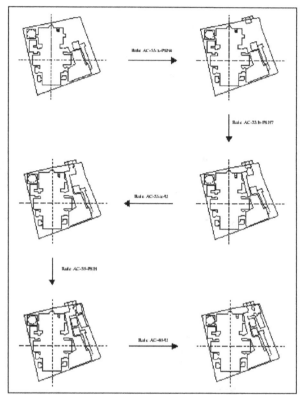

Fig. 11b

8. Generating the overall layout of al-Ashraf Barsbay Madrasa after applying the rules for determining the thicknesses of walls by applying the rules from (WT-2, WT-5, WT-6, WT-7, WT-8, and WT-9) (fig. 12a):

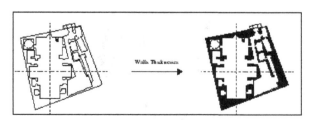

Fig. 12a

9. Assigning openings for Qibla Iwan by applying the rules OP-1, OP-2, and OP-7 (fig. 12b):

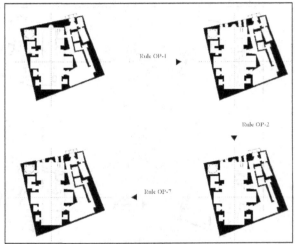

Fig. 12b

10. Assigning openings for tomb by applying the rules OP-8 and OP-9 (fig. 12c):

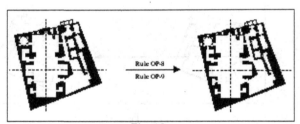

Fig. 12c

As a result of the ten-stage derivation process the floor plan of al-Ashraf Barsbay madrasa is produced. fig. 13 illustrates the floor plan after the rule applications in the ten stages:

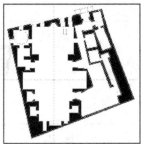

Fig. 13

6 Conclusion

In this paper, a new framework that systematically describes the morphological structure and the derivation process of Mamluk madrasas is introduced. The framework represents an explicit and externalized process for generating the forms of architectural precedents. Within this framework, a formal language for the Mamluk madrasa floor plans is formulated. A set of vocabulary elements is concluded from a deductive morphological analysis of a group of study cases. The study sample includes sixteen madrasas selected from Egypt, Syria, and Palestine. These Mamluk madrasas share common features and vocabularies in plan composition. The features include location of components such as courtyard, Iwans, tomb, sabil and minarets; exterior layout shape, orientation and topological relationships. Visual aspects such as proportion, symmetry, axial organization, and rotation are also shared by the study sample instances.

A set of grammatical rules is formulated to guide the sequential process of composing the vocabulary components into a meaningful plan that can be considered a member of the family of Mamluk architectural styles. The grammar set consists of ninety-three rules for a step-by-step derivation of a plan of Mamluk madrasas. The process of generating a floor plan for a Mamluk madrasa involves applying rules in many phases. These include rules for generating schematic external layout shapes, architectural components, interior spaces, Iwans and courtyards, spaces between Iwans, lateral spaces, mihrab in the Qibla-Iwan, Tomb or Mausoleum, ablution spaces, cells and rooms around the courtyard, the main entrance, derka and openings. In addition, there are rules for: determining the orientation of spaces, scaling Iwans and courtyards, articulating interior spaces, determining wall thickness and, finally, the termination rules.

In order to develop the formulated formal language, this research adopted two main methods: the case study approach (which focused on the educational buildings in Mamluk architecture); and the theoretical research method (represented by the deductive analysis and the morphological study, as well as the mathematical and the linguistic representations for developing shape grammars).

The theoretical framework is mathematically and geometrically developed to explore new layers in the design of beautiful Mamluk buildings. Possible future extensions may focus on a morphological study and design of a generative system for the facades, the three-dimensional massing organizations, the decoration articulation or the opening treatment of Mamluk buildings. Further, our research has been concerned with the development of the formulated framework into computer-aided precedent-based educational software that trains students to discover new aspects and layers of existing Mamluk architectural examples and explore emergent designs that can be considered new imaginary additions to the same architectural style applied possibly to recent functions [Eilouti and Al-Jokhadar 2007].

References

AL-JOKHADAR, A. 2004. Shape Grammars: An Analytical Study of Architectural Composition Using Algorithms and Computer Formalism, Unpublished Master Thesis, Jordan University of Science and Technology, Jordan.

BIANCA, S. 2000. *Urban Form in the Arab World: Past and Present.* London: Thames & Hudson Ltd.

BURGOYNE M.H. 1987. *Mamluk Jerusalem: An Architectural Study.* Published on behalf of the British school of archeology in Jerusalem by the Islam Festival Trust, London.

CHOMSKY, N. 1957. *Syntactic Structures.* The Hague: Mouton and Co.

DING L., GERO J.S. 2001. The emergence of the representation of style in design. *Environment and Planning 'B': Planning and Design* **28**: 707- 731.

EILOUTI, B. 2003. Three-dimensional modeling by replacement. In: *Proceedings of the IASTED International Conference: Modeling, Simulation, and Optimization* (July), Banff, Alberta, Canada.

EILOUTI, B. and A. ALJOKHADAR. 2007. A Computer-Aided Rule-Based Mamluk madrasa Plan Generator. *Nexus Network Journal* **9**, 1: 31-58.

GIPS J. 1975. *Shape Grammars and their Uses.* Basel: Birkhäuser.

GLASSIE, H. 1975. *A Structural Analysis of Historical Architecture.* Knoxville: University of Tennessee Press.

HIMMO, B. 1995. *Geometry Working out Mamluk Architectural Designs: Case Study: Sabil Qaytbay in Holy Jerusalem.* Master Thesis, University of Jordan, Jordan.

The Islamic Methodology for the Architectural and Urban Design. 1991. Proceedings of the 4th seminar, Rabat. Morocco Organization of Islamic Capitals and Cities.

KNIGHT T.W. 1999. Shape grammars: six types. *Environment and Planning 'B': Planning and Design* **26**: 15-31.

MARCH L. 1999. Architectonics of proportion: historical and mathematical grounds. *Environment and Planning 'B': Planning and Design* **26**: 447-454.

MICHEL, G., 1996. *Architecture of the Islamic World: Its History and Social Meaning.* London: Thames and Hudson.

MITCHELL W.J. 1986. Formal representations: a foundation of computer aided architectural design. *Environment and Planning 'B': Planning and Design* **13**: 33-162.

OSMAN M.S. 1998. Shape grammars: simplicity to complexity. Paper presented in University of East London. (http://www.ceca.uel.ac.uk/cad/student_work/msc/ian/shape.html, accessed 1 Oct 2003).

Organization of Islamic Capitals and Cities. 1990. *Workshop of Architectural and Urban Design Fundamentals in Islamic Periods,* Islamic Architecture Heritage Conservation Center, Cairo, Egypt 1990.

PARKER, R. 1985. *Islamic Monuments in Cairo.* 3rd ed. Cairo: The American University in Cairo Press.

SERAGELDIN, R. 1988. Educational Workshop. In *Places of Public Gathering in Islam.* Proceedings of Seminar Five, Architectural Transformations in the Islamic World. Amman, Jordan: The Aga Khan Foundation for Architecture.

STEADMAN J.P. 1983. *Architectural Morphology: An Introduction to the Geometry of Building Plans.* London: Pion.

STIERLIN, Henri 1984. *Great Civilizations: The Cultural History of the Arabs.* London: Aurum Press.
———. 1996. *Islam: Early Architecture from Baghdad to Cordoba.* Taschen Publishers.

STINY G. and MITCHELL W. J. 1980. The grammar of paradise: on the generation of Mughul gardens. *Environment and Planning 'B': Planning and Design* **7**: 209-226.

STINY G. 1980. Introduction to shape and shape grammars. *Environment and Planning 'B': Planning and Design* **7**: 343-351.

WILLIAMS, John Alden. 1984. Urbanization and Monument Construction in Mamluk Cairo. In *Muqarnas II: An Annual on Islamic Art and Architecture.* Oleg Garbar (ed). New Haven: Yale University Press.

About the authors

Buthayna H. Eilouti is Assistant Professor and Assistant Dean at the Faculty of Engineering in Jordan University of Science and Technology. She earned a Ph.D., M.Sc. and M.Arch. degrees in Architecture from the University of Michigan, Ann Arbor, USA. Her research interests include Computer Applications in Architecture, Design Mathematics and Computing, Design Pedagogy, Visual Studies, Shape Grammar, Information Visualization, and Islamic Architecture.

Amer Al-Jokhadar is a Part-Time Lecturer at the College of Architecture and Design in German-Jordanian University in Amman, and Architect in TURATH: Architectural Design Office. He earned M.Sc. and B.Sc. in Architectural Engineering from Jordan University of Science and Technology. His research interests include: Shape Grammar, Mathematics of Architecture, Computer Applications in Architecture, Islamic Architecture, Heritage Conservation and Management, Design Methods.

Buthayna H. Eilouti

Department of Architectural Engineering
Jordan University of Science and
Technology
POB 3030
Irbid 22110, JORDAN
buthayna@umich.edu

Amer M. I. Al-Jokhadar

College of Architecture & Design
German-Jordanian University
Amman, JORDAN

TURATH: Heritage Conservation
Management and Environmental Design
Consultants
P.O.Box 402, Amman, 11118 JORDAN
amerjokh@hotmail.com
amerjokh@gmail.com

Research

A Computer-Aided Rule-Based Mamluk Madrasa Plan Generator

Abstract. A computer-aided rule-based framework that restructures the unstructured information embedded in precedent designs is introduced. Based on a deductive analysis of a corpus of sixteen case studies from Mamluk architecture, the framework is represented as a generative system that establishes systematic links between the form of a case study, its visual properties, its composition syntax and the processes underlying its design. The system thus formulated contributes to the areas of design research and practice with a theoretical construct about design logic, an interactive computerized plan generator and a combination of a top-down approach for case study analysis and a bottom-up methodology for the derivation of artifacts.

Keywords: Computer-aided design, logic of design, visual reasoning, precedent-based design, design process, mamluk architecture, shape grammar

1 Introduction

Design involves multiple activities and tasks that are based on reusing past design solutions. Design knowledge gained from studying precedents plays a significant role in the pre-design and precept reasoning stages. This is due to the fact that previous experiences help in understanding new situations and in casting older solutions into new problems, at least in the early stages of the problem-solving process. Precedent-based design experience is used when performing different tasks, whether they involve routine activities or require creative contributions. A process of interpolation and matching is conducted for the routine activities of comparison and exclusion in the processing of the creative ones. This is also the case in architectural design, where many problem-solving tasks are based on modifying past solutions. In fact, a major part of the pre-design investigation in architecture is concerned with studying case studies (precedents) that are related to the problem at hand. Multiple layers of information can be inferred from case studies to form a point of departure for new designs.

Starting with a precedent-based design model can enhance the solution of a new design problem by taking as a point of departure a case as a whole or a combination of selected parts or features of different cases. In order to be of practical use, information inferred from precedents has to be represented in a meaningful way so as to enable designers to restructure and reassemble the extracted data for the generation of a new design. Precedent-based models can be represented in many forms. The one that will be emphasized in this paper is the Rule-Based Design (RBD) model, which Tzonis and White [1994, 20] called the Principle-Based Reasoning model. RBD provides dynamic and flexible solutions since many alternatives can be derived by applying a combination of rules (selected from a finite set) on a small set of basic shapes. In RBDs, data extracted from a group of precedents is formulated as concise productive rules and defined vocabulary elements by using a top-down approach. In this approach, a design problem is studied as a whole unit to derive the basic constituent elements and relationships underlying its structure. In this regard, RDB models can be associated with Problem-Based Learning techniques [Dochy et al. 2005] to strengthen their pedagogical power. In Problem-Based Learning settings, a whole problem is introduced, investigated and analyzed to allow learners to derive all the relevant basic data and details. For scope limitations, issues of Problem-Based Learning associations and impacts will be discussed in a separate paper.

The main hypothesis of this study is that the logic of form making and style definition can be represented as a finite set of generative design rules and shape compositional principles. Thus, information embedded in precedents can be represented by a finite set of planning procedural rules and parameters for aesthetic articulation.

The main scope of this research is to develop a computer-aided precedent-based system that can describe, analyze and generate two-dimensional representations of architectural design by applying productive rules of composition. The system is implemented on the floor plans of educational buildings (*madrasas*) in Mamluk architecture as case studies.

The main goal of the study is to formulate a structured and systematic framework for analyzing the form of historical artifacts and for generating emergent designs based on the morphological analysis of precedents. The main objective is to restructure the originally unstructured or weakly-structured information embedded in precedents in the form of clearly formulated rules to make the information more usable and recyclable.

The rule-based design framework developed in this paper is expected to contribute to the theory, practice and teaching of design. Based on a structuralist view[1] of the morphology of precedents [Caws 1988], the framework proposes a theoretical construct that encompasses a system of compositional and procedural rules for form generation. It enhances the body of knowledge in design areas by means of an externalized and explicit method of form-making and an incremental process for architectural composition, and by an applicable system for restructuring the weakly-structured information that is embedded in precedent designs. Furthermore, the paper emphasizes the aesthetic values of Mamluk architecture by systematically analyzing its morphology. For design practice, within the scope of this research, the use of the framework can help designers to visualize and evaluate several solutions of a given design problem, particularly in two-dimensional representations, and then to select one or a combination of these designs to develop. In addition, designers can automatically and systematically generate designs based on a knowledge base developed from the study of selected precedents. Regarding the pedagogical significance, applications in analysis and synthesis of basic components of compositions have important implications for design instructors who want to communicate the principles of visual composition and guidelines for the design process. In addition, the proposed framework represents an efficient way of learning about styles or languages of designs, especially about their compositional aspects. A language paradigm for form generation can also reveal general design strategies that students can learn from and use in solving their own design problems. By applying a bottom-up approach of rule search and matching to initial shapes, a student may better understand the form-making process and its associations with style definition.

The research introduces a framework for a set of case studies, the morphology of which did not receive enough attention in the areas of the systematic analysis and synthesis, and the explicit articulation of the processes of their form-making. The precedents are selected from Mamluk architecture and are widely considered beautiful representative exemplars of Islamic architecture. The formulated framework proposes an integrated system that consists of a multi-layer structure and multi-phase sequence for deriving floor plans of Mamluk architecture.

2 Background

Most previous efforts in the area where precedent-based and rule-based design studies intersect are represented in the form of shape grammars. Shape grammar [Gips 1975; Stiny 1980, 1994; Flemming 1987] is a method of the description, analysis, representation, classification, and generation of the form of an artifact and its incremental design process. In most cases, shape grammars are associated with the analysis of precedents. The common factor in the precedents studied is usually their designer, their style or the similar temporal or geographical context in which they were designed. Design precedents are typically investigated in order to understand the geometrical composition of their appearance and the process of their production. Then, a group of possible rules and initial shapes is derived.

There are two principle kinds of shape grammar: *standard grammar* and *parametric grammar*. While in standard grammars most attributes of shapes are constant, in parametric grammars they tend to be more flexible and variable. Because of its flexibility, parametric grammar is more practical and popular than the standard one. If the number of parameters in parametric grammars is small, designers can recognize and predict several options of designs, but if the number is large and the spatial relations are more complex, shape grammars become too complex to be applied manually. Computer implementations of shape grammars facilitates the search through complex grammars or large numbers of rules and parameters. The relationship between shape grammars and computer implementations can be described in the statement:

> Shape grammars naturally lend themselves to computer implementations: the computer handles the bookkeeping tasks (the representation and computer computation of shapes, rules, and grammars, and the presentation of correct design alternatives) and the designer specifies, explores, develops design languages, and selects alternatives [Tapia 1999, 1].

The concept of being able to derive new instances of a given style is certainly interesting as a way of extending the scope of a style by adding imaginary members that have the same compositional attributes. Such a derivation is based on the understanding of the formal patterns and compositional principles underlying the appearance of designed objects that belongs to that style. Consequently, it is fundamental to formulate the rules of a grammar in a flexible way so that the unexpected elements of design are allowed to emerge.

In order to automate and implement shape grammars in an exploratory way, it is essential to formulate rules in an algorithmic, parametric and interactive format so when they are fed into the computer, they generate multiple alternatives of a design. The derived designs present possibilities that may inspire the designer. The rules should be based on the conceptual as well as the formal components of a design, and on the relational, morphological, and topological associations among all components. Computer implementations of shape grammars can function as educational tools for demonstrating the range and power of the grammars, and for illustrating the morphology and process of designs. They enable students and designers to generate, analyze, evaluate, and select more rapidly and easily from among the various design alternatives that grammars generate without having to deal with the technicalities of grammar development.

In spite of their theoretical appeal, shape grammars did not receive enough attention in the area of their automation. This is most probably the result of the relative complexity of their underlying algorithms and the difficulty of developing an integrated system for shape analysis and synthesis [Tapia 1999].

According to Gips [2000], it is possible to identify four major types of the computer implementation of shape grammars. These are:

1. **A shape interpreter.** In this type, a user defines a shape grammar in the computer, and the program generates shapes in the given language, either automatically or guided interactively by the user.

2. **A parsing program.** A program of this type is given a shape grammar and a shape. The program determines if the shape belongs to the language generated by the grammar and, if so, gives the sequence of rules that produces the shape. This type focuses on the analysis and search aspects rather than the design generation issues.

3. **An inference program.** In this type, the program is fed a coherent set of shapes of a given style, for which the computer automatically generates a shape grammar for the same style.

4. **A computer-aided shape grammar design program.** A program of this type assists users in designing a shape grammar by providing sophisticated tools for rule construction and development.

Some examples of the implementations that have been conducted on shape grammars are listed in Table 1. Gips's interpreter [1975] enables the user to input a simple two-rule shape grammar; the program then generates two-dimensional shapes. Only polygonal vocabulary elements are allowed in this grammar. The program ignores the issue of detecting sub-shapes. Krishnamurti [1980, 1981-a, 1981-b] did pioneering work on developing data structures and algorithms for solving the sub-shape transformation and rule application problems for two-dimensional grammars. Using the Prolog programming language, Flemming [1987] developed a grammar for three-dimensional representations of the Queen Anne house, and Chase [1989] developed a general two-dimensional shape grammar system. Tapia's GEdit [1999] is a general two-dimensional shape grammar interpreter that supports sub-shape detection and shape emergence. Agarwal and Cagan's coffee maker grammar [1998] has been implemented using a JAVA program. A shape grammar interpreter was designed by Piazzalunga and Fitzhorn [1998] for three-dimensional oblong manipulations using the LISP language. Also using a version of LISP, AutoLISP, Eilouti's FormPro1 [2001] develops a universal grammar for three-dimensional architectural compositions with possible style variations for each resultant design. Duarte's program [2005] implements a shape grammar for Malagueria house designs.

In most previous efforts, computational problems related to encoding rules and their executions were focused on producing abstract shapes rather than architectural representations. They considered the layer of form processing that mostly concentrates on the construction grid and overall shapes of actual designs. Classification of components according to their functional and architectural articulations was in most cases ignored.

Furthermore, in most efforts, issues of designing the interactive and user-friendly interface of a program did not receive enough attention. Thus, most of these systems were not easy to use for nonprogrammers, novice users of shape grammars, or design practitioners.

1	Simple interpreter	Gips, 1975	SAIL [Standard Artificial Intelligence Language]
2	Shape grammar interpreter	Krishnamurti, 1981-a and 1981-b	
3	Shape generation system	Krishnamurti and Giraud, 1986	PROLOG
4	Queen Anne houses	Flemming, 1987	PROLOG
5	Shape grammar system	Chase, 1989	PROLOG / Mac
6	GEdit	Tapia, 1999	LISP / Mac [Macintosh Common LISP (MCL)]
7	Shape grammar interpreter	Piazzalunga and Fitzhorn, 1998	ACIS/LISP
8	Coffee maker grammar	Agarwal and Cagan, 1998	JAVA
9	FormPro1	Edouti, 2001	AutoLISP
10	Malagueria Program	Duarte, 2005	

Table 1. A List of shape grammar computer implementations

The computer-aided generative system presented in this paper is a multi-layer, multi-phase shape interpreter that transforms a layout construction grid derived by rule applications in one phase into a full architectural plan with walls and openings in another phase. It is designed with a quickly-learned and easily-used interface. Furthermore, the stylistic prototypes of the program outcomes are developed from a valuable set of precedents that have not received enough attention of researchers, especially in the systematic and applied research areas. This set consists of Mamluk educational precedents.

3 The theoretical framework for the rule-based generative system

Grammatical inference can be defined as the task of logically inferring a finite set of rules from the systematic investigation of a coherent set of designed objects. In its architectural application, a building design or group of designs is defined. The designs are analyzed by decomposing them into a vocabulary, which represents the lexical level, and rules, which represent the syntactic level of their structure. Spatial relations, or arrangements of vocabulary elements in space, are identified in terms of the decompositions of the original designs. As a result of the re-synthesis of the decomposed elements, the shape grammar generates the original designs and possible new designs that did not exist before. New designs are generated by restructuring new combinations of the vocabulary elements in accordance with the inferred spatial relations, or by re-assembling the same elements using different rules or sequences.

In order to develop a generative design system which applies rule-based assembly models, a foundation of a flexible parametric representation of rules and basic shapes must be formulated. The rules and initial shapes are inferred from a set of precedents, drawn in the present paper from Mamluk architecture. The study focuses on a set of educational building (*madrasas*) that belong to this style. Mamluk madrasas represent some of the most beautiful examples of the Islamic architecture. Despite their historical and aesthetic value, they have not received enough attention from researchers, especially in the areas of systematic analysis of their morphological structures or designs.

This research examines a group of sixteen Mamluk madrasas from the different regions ruled during the Mamluk period. The case studies include thirteen precedents from Egypt (located in Cairo), one precedent from Syria (located in Aleppo), and two precedents from Palestine (located in Jerusalem). The study cases from Cairo are: al-Ashraf Barsbay, al-Kadi Zein al-Dien Yehya, al-Amir Kerkamas, al-Sultan Inal, al-Ghouri, Umm al-Sultan Sha'ban, Kani Bay Kara al-Remah, al-Sultan Kaytebay, and Abu-Baker Mezher Madrasas from the covered court type; and al-Thaher Barquq, al-Sultan Hasan, Serghtemetsh, and al-Sultan Qalawun Madrasas of the open court type. The study cases from Jerusalem are: the Tashtamar Madrasa from the covered court type and al-Baladiyya Madrasa of the open court type. The case from Aleppo is al-Saffaheyya Madrasa which is of the open court type. The morphological study of the case studies, as well as the spatial organization of the components is discussed in a separate paper [Eilouti and Al-Jokhadar 2007].

As a result of the structuralist analysis (see note 1) to find the commonalities that underlie the cases, and the comparisons of the sixteen precedents, it is concluded that in order to reconstruct a floor plan of a Mamluk madrasa, five sets of rules need to be defined. The first set is the rules for schematic layout. Applied in a top-down, out-in approach where form evolves from the exterior-most abstracted shapes into the interior and detailed articulations, rules in this set are used to derive the shape of the exterior layout and its major organizational axes. The second set consists of the rules for the architectural components, which articulates the interior space and room shapes based on their functions and the logic of how the components are related within the floor plan diagram. Rules in the first two sets are schematic and are applied to establish the construction lines and major axes of orientation and symmetry. The third set includes rules for determining wall thicknesses. These assign a thickness for the construction lines generated through the layout and architectural component rules. Thus, rules in the third set transform the schematic diagram generated by the first two sets into an architectural representation of the floor plan at hand. The fourth set consists of rules for openings. These add shapes for door and window openings to the floor plan layout generated by the previous three sets of rules. The fifth set includes the rules for termination. These conclude the process of floor plan generation.

The five sets of rules and the information they represent are illustrated in Table 2. Each set represents a phase in the derivation sequence. The generation process starts with phase one, cycles between the second, third and fourth phases as needed, and finally concludes with the fifth phase, which terminates steps of the process. Rules in each phase are listed with their count and information as follows:

The formulated grammar consists of 93 rules distributed as:

Rule type	Number of Rules
Rules for generating schematic external layout shapes	12
Rules for generating architectural components	59
Rules for determining wall thickness	9
Rules for generating openings	9
Termination rules	4

The rules in each category are listed in Table 2.

Rule Phase	Rule Type	Number of Rules	Information Represented
First Phase: Exterior Shape	**Rules for generating schematic external layout shapes**	12	**Exterior shape geometry, proportions, orientation**
Second Phase: Architectural Components	Rules for generating interior spaces	2	Shape, location, proportion, topological relations
	Rules for determining the orientation of spaces	1	Angles and location of shape lines
	Rules for generating Iwans & courtyards	3	Location and division into subshapes
	Rules for sealing Iwans & courtyards	3	Axes, divisors and proportion
	Rules for generating spaces between Iwans	2	Location of transitional spaces, proportion and dimensions
	Rules for generating lateral spaces	4	Shape, dimensions and set operations
	Rules for generating Mihrab on the Qibla-Iwan	3	Axis of symmetry, location of semi-circle, transformation, set operation
	Rules for articulating interior spaces	4	Count type, roof type, location and geometry of fountain, sadla & minbar, iwan divisions
	Rules for generating Tomb or Mausoleum	13	Location, proportion, orientation, scale, shape and topological relations
	Rules for generating Ablution spaces	2	Location, orientation, shape and topological relations
	Rules for generating cells & rooms around the courtyard	4	Shape, transformations and topological relations
	Rules for generating the main entrance and Derka	18	Shape, location, proportion, size and topological relations
Third Phase: Wall Thickness	**Rules for determining wall thickness**	9	**Wall thickness**
Fourth Phase: Openings	Rules for generating openings	9	Placement, size and set operations
Fifth Phase: Process Termination	**Termination rules**	4	**Removal of construction lines and labels**

Table 2: The rule system for the derivation of Mamluk madrasa precedents

The Mamluk madrasa grammatical rule numbering system, and the initial and the exterior layout shapes are illustrated a separate paper [Eilouti and Al-Jokhadar 2007].

As an example of how the formulated rules restructure the information embedded in the case studies, rules (AC-5-D) and (AC-6-M) specify that there are transitional spaces between the courtyard and two Iwans. In addition, Rule (AC-7-S) specifies the dimensions of the courtyard (V_B) and (H_B). The proportion between (V_B) and (H_B) is 1:1, while the proportion between the courtyard and the total interior space is around 1:2.54, 1:2.30, or 1:2.22.

Rules (AC-8-S) and (AC-9-S) determine the dimensions of the two Iwans ($H_C : V_C$) and ($H_D : V_D$). All of these spaces are symmetrical about the main axis $Y_2 - Y_2$. Fig. 1 illustrates three of these rules and their associated symbols. These rules, namely (AC-7-S), (AC-8-S) and (AC-9-S), are used for scaling two of the major components of madrasa plan morphology, the courtyard and the Iwan components.

As an example of the information extracted from the floor plans of the sixteen precedents, the mathematical relations and ratios that govern the courtyard geometry of the case studies are illustrated in Table 3. The H and V values in the table denotes the horizontal and vertical dimensions of the court B. B represents the area of the court, and A represents the area of the court and the spine of the Iwans. B:All represents the area of the court compared to the total area of the madrasa at hand. X,Y signifies the origin point which is usually the centre of courtyard. These proportional and locational values of shape are translated into parametric assignment for the rules in the generative system structure.

A numerical analysis similar to that illustrated in Table 3 is conducted on all architectural components of the case studies.

The five-fold framework outlined above is developed into a computer implementation, which is designed to be viable for practical design problems. The computerized framework is designed to enable designers to understand existing designs and to generate new forms that have the same style of the studied precedents.

The computerized version of the generative system reflects the multi-layered multi-phased characteristics of the rule structure in order to permit a rapid generation of Mamluk madrasa prototypes through an incremental and lucid step-by-step sequence of selecting shapes, operators, transformations, parameter assignment, and rule application.

The method used to develop the generative system for the Mamluk madrasa design can be applied to any set of precedents and can be extended to allow experimentation and exchange between designs and styles.

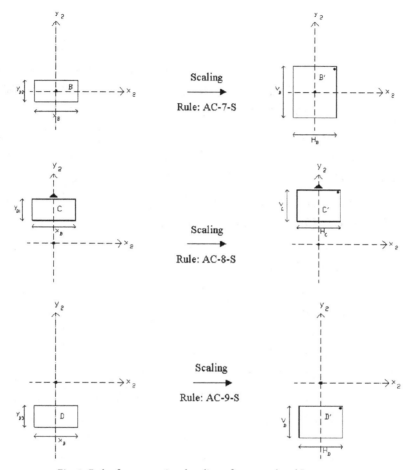

Fig. 1. Rules for proportional scaling of courtyard and Iwans

No.	Case Study	B					
		Courtyard (al-Sahn or dorqa'a)					
		H₃	V₃	H₃ : V₃	B : A	B : All	(x, y)
1	al-Ashraf Barsbay Madrasah	14.35	16.00	1 : 1.11	1 : 2.30	1 : 6.7	(0,0)
2	al-Kadi Zein al-Din Yehya Madrasah	7.70	7.50	1 : 1.02	1 : 2.30	1 : 11	(0,0)
3	al-Amir Karkamas Madrasah	8.60	8.60	1 : 1	1 : 2.57	1 : 4.45	(0,0)
4	al-Sultan Inal Madrasah	10.50	10.50	1 : 1	1 : 2.54	1 : 5	(0,0)
5	al-Ghuri Madrasah	9.50	9.90	1 : 1.02	1 : 2.66	1 : 9.4	(0,0)
6	Umm al-Sultan Sha'aban Madrasah	11.20	13.90	1 : 1.22	1 : 2.30	1 : 6	(0,0)
7	Kane Bay Kara al-Remah Madrasah	7.40	7.60	1 : 1.02	1 : 2.54	1 : 8.5	(0,0)
8	al-Sultan Katebay Madrasah	8.65	8.65	1 : 1	1 : 2.52	1 : 12	(0,0)
9	Abu Baker Mezher Madrasah	9.55	7.90	1 : 1.22	1 : 2.41	1 : 5.3	(0,0)
14	al-Tashtamur Madrasah	30.25	29.20	1 : 1.02	1 : 2.37	1 : 6.5	(0,0)
10	al-Thaher Barquq Madrasah	17.90	22.00	1 : 1.22	1 : 2.22	1 : 7	(0,0)
11	al-Sultan Hasan Madrasah	31.40	35.00	1 : 1.11	1 : 2.15	1 : 6.55	(0,0)
12	al-Amir Derghtemsh Madrasah	21.18	17.12	1 : 1.22	1 : 2.15	1 : 5	(0,0)
13	al-Sultan Qalawun Madrasah	15.85	19.75	1 : 1.22	1 : 2.44	1 : 3.95	(0,0)
15	al-Saffaheya Madrasah	8.60	7.12	1 : 1.22	1 : 2.22	1 : 6	(0,0)
16	al-Baladiyya Madrasa	13.20	12.95	1 : 1.02	1 : 2.30	1 : 5.5	(0,0)

Table 3. The parameters that control the shapes of the courtyards of Mamluk madrasas

4 The program interface

In general, it is widely assumed that people who use shape grammars tend to be visual thinkers, and those who program and implement computer codes are symbolic thinkers. People who think well both visually and symbolically seem be quite rare [Gips 2000]. This may explain the insufficiency of computer implementations of shape grammar systems and the problem of the interface design in the existing implementations.

The issue of successful user design interface is significant in the design of computer implementations for shape grammars. Interface issues were addressed in a comprehensive way in Tapia [1996; 1999]. Tapia emphasizes the importance of improved computational machinery for a general two-dimensional shape grammar interpreter, along with a simple, intuitive, visual interface. The user of a computer-aided rule-based interpreter program needs to learn both about rule structure and how to use the program. To facilitate this, the program design should be transparent. Certainly it should make it easier to use the program than to try out rule application by hand [Gips 2000].

To generate or reconstruct a floor plan for Mamluk madrasas, the program starts by parsing and classifying the main vocabulary elements, their relationships and possible transformations. The overall flowchart of the rule development system is shown in fig. 2. Because of the large number of rules, the program commands are designed in phases and are decomposed into sequences of rules. The codes of the generative system are written in the AutoLISP language for the AutoCAD operating environment. This saves time in rewriting some existing programming routines and is very suitable for graphic applications.

The interface of the program described here consists of five main menus in addition to the typical file management, editing, viewing and support window-compatible menus. The five menus represent the five phases of the derivation process described above. Each phase reveals the multi-layered structure of the madrasa composition at that stage. The menus are designed in a simple and clear format. The menus are:

1. "Schematic Layout" rules. This menu specifies seven functions: the first six illustrate the six different types of layout for generating the exterior layout shapes of madrasa floor plans. Each layout has two rules. The first determines the proportions of the sides of the shape, and the second determines the angles between each two sides. The seventh and last function in this menu defines the centre of the exterior layout with reference to the interior court.

2. "Architectural Components" rules. This menu has nine submenus:

    (1) Interior spaces. This submenu includes three rules for inner space organization.

    (2) Courtyard and Iwans. This submenu includes twelve rules for the open and semi-open space articulation within the exterior layout shape.

    (3) *Mihrab*. This submenu includes three rules for locating the niche in the southern Iwan wall.

    (4) Articulating interior spaces. This has three rules for shaping the minor inner spaces.

    (5) *Sadla*. This submenu includes one rule for forming the geometry of the secondary Iwan.

    (6) Tomb. This submenu branches into four submenus: tomb #1 (with four rules); tomb #2 (with four rules); tomb #3 (with four rules); and tomb #4 (with one rule).

    (7) Ablution space. This submenu includes two rules for the articulation of the ablution room.

    (8) Cells around the courtyard. This submenu includes four rules for the articulation of the supportive cell geometry.

    (9) Main entrance and *derka*. This submenu branches into six submenus: entrance #1 (with three rules); entrance #2 (with three rules); entrance #3 (with three rules); entrance #4 (with three rules); entrance #5 (with three rules); and entrance #6 (with three rules).

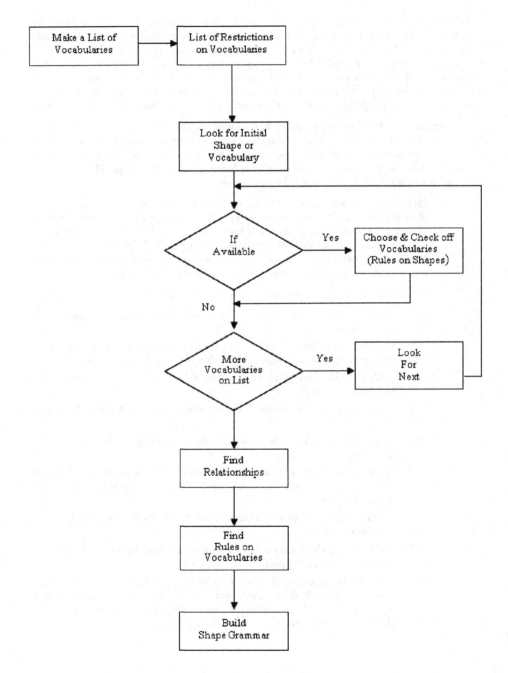

Fig. 2. A flow chart illustrating the process of developing the rule-based model

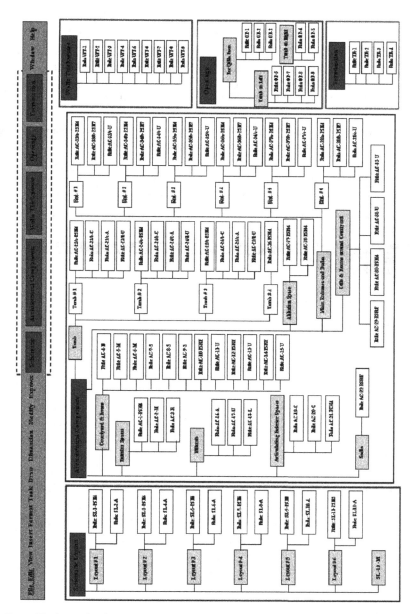

Fig. 3. The hierarchical organization of the rules of the Mamluk Madrasa program interface

3. "Walls Thicknesses" rules. The main function for this menu is determining the thicknesses of exterior and interior walls of madrasas. For applying thickness to designated walls, the menu offers nine rules.

4. "Openings" rules. The main function for this menu is assigning the doors and windows for the plan of the madrasa. This is done through specifying whether the tomb is located on the right or left side of the Qibla-Iwan. There are nine rules under this menu.

5. "Termination" rules. These will erase the labels assigned to shapes through different stages for developing the shape grammars. This menu has four rules.

All of the five main menus and the additional submenus are illustrated in the hierarchical organization diagram shown in fig. 3. The information described by the rules is listed in Table 2.

The structure of the aforementioned five menus is shown in images captured from the screen, as illustrated in figs. 4 to 8.

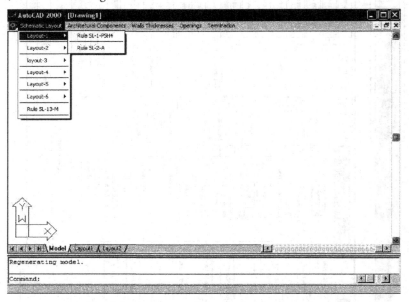

Fig. 4. A screen image of the schematic layout menu

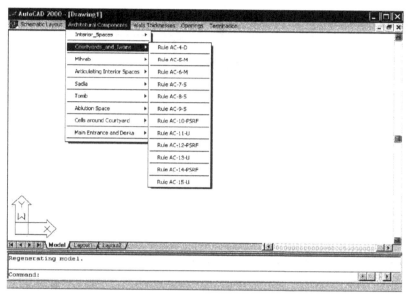

Fig. 5. A screen image of the architectural component menu

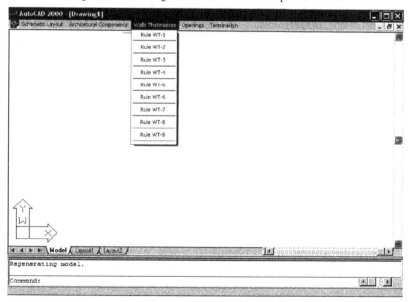

Fig. 6. A screen image of the wall thickness menu

Fig. 7. A screen image of the openings menu

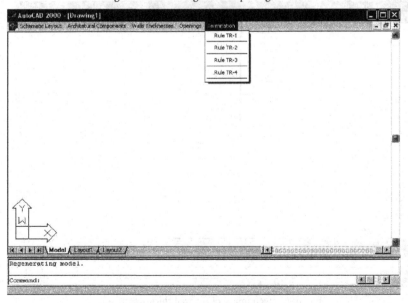

Fig. 8. A screen image of the termination menu

5 Program testing and discussion

The Case of al-Ashraf Barsbay Madrasa in Cairo is used to demonstrate the applicability of the program presented in the previous sections. This case represents one of the sixteen precedents analyzed. It exhibits most of the features and components of the generative system. The generation process starts with the basic layout shape shown in fig. 9. The shape of the exterior layout is selected from the set of initial shapes offered by the schematic layout menu.

The plan generation process continues by applying rules from the architectural component menu to add the overall layout shape of the interior spaces and to assign numerical values for the parameters required by the rule prompts. The result of the interior space rule application process is illustrated in fig. 10.

Other architectural components, such as the courtyards, *iwans*, *mihrab*, *minbar*, *sadla*, tomb, ablution space, cells around the courtyard, the main entrance, *derka* and the Sheikh's house, are also added by using the architectural components menu. This set of architectural component configuration and articulation steps of the derivation process is illustrated in figs 11-17.

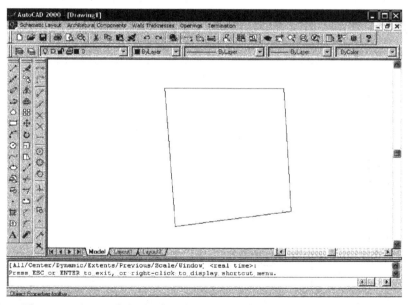

Fig. 9. The exterior layout of al-Ashraf Barsbay Madrasa after applying rules from the schematic layout menu

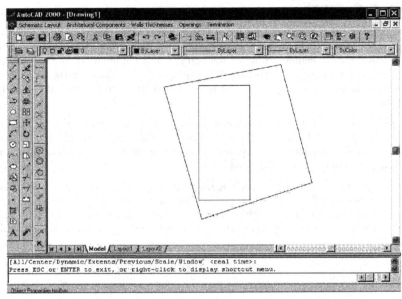

Fig. 10. The overall layout of the interior spaces of al-Ashraf Barsbay Madrasa

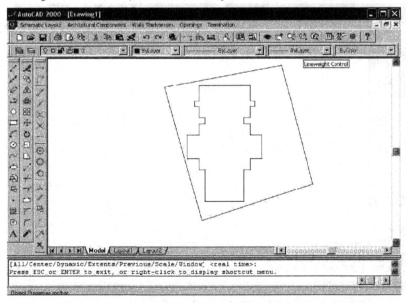

Fig. 11. The addition of the courtyards and *iwans* for al-Ashraf Barsbay Madrasa

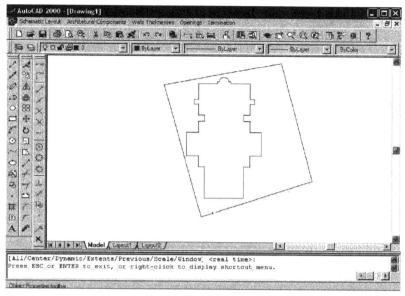

Fig. 12. Generating the *mihrab* for al-Ashraf Barsbay Madrasa

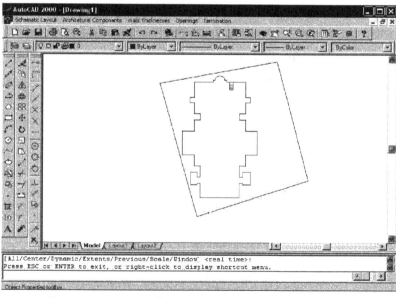

Fig. 13. Generating *minbar* and *sadla* for al-Ashraf Barsbay Madrasa

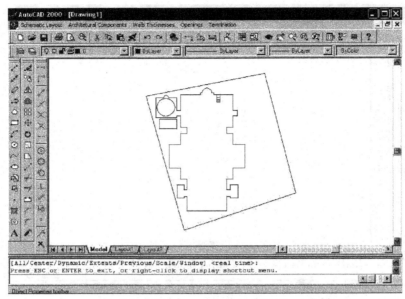

Fig. 14. Generating the tomb for al-Ashraf Barsbay Madrasa

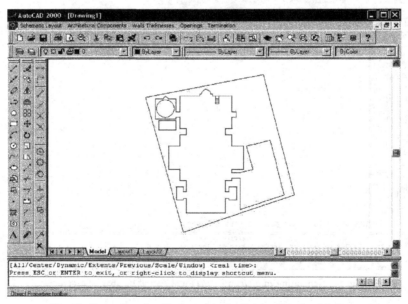

Fig. 15. Generating the ablution space for al-Ashraf Barsbay Madrasa

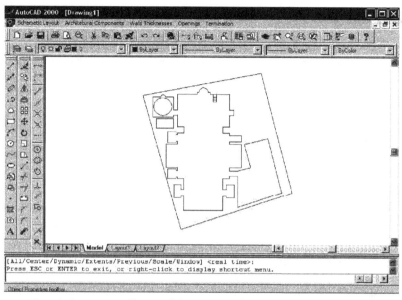

Fig. 16. Generating cells around the courtyard for al-Ashraf Barsbay Madrasa

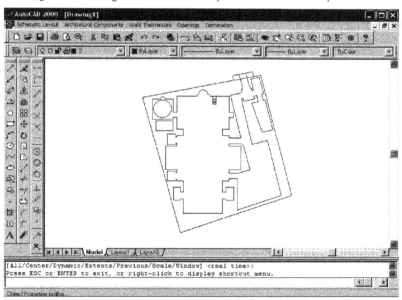

Fig. 17. Generating the main entrance, *derka*, and the Sheikh's house for al-Ashraf Barsbay Madrasa

After the definition of the main architectural components of the madrasa, the door and window openings are added by using the opening menu. The plan with the openings is shown in Fig. 18.

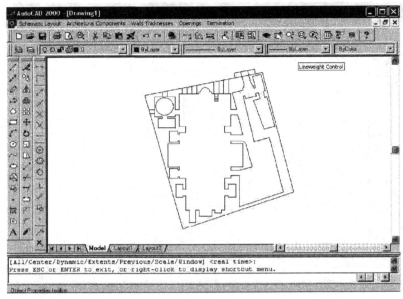

Fig. 18. Generating openings (doors and windows) for al-Ashraf Barsbay Madrasa

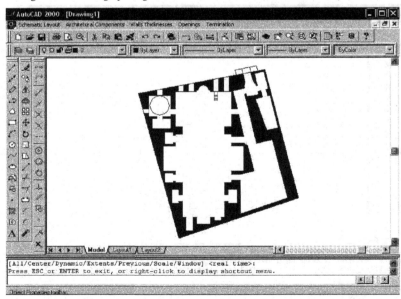

Fig. 19. The overall layout of al-Ashraf Barsbay Madrasa after applying rules for the assignment of wall thicknesses

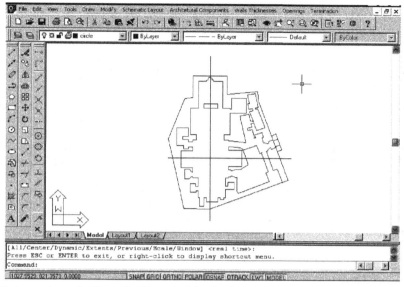

Fig. 20. A new alternative for the final layout of al-Ashraf Barsbay Madrasa after changing the exterior layout, location of the entrance, *derka*, and the tomb (in front of *qibla iwan*)

The penultimate step of the derivation process is the assignment of wall thicknesses to the lines derived so far. The assignment is conducted by applying the wall thickness menu commands. The result of this process, which transforms the construction line grid into the representation of the architectural plan, is shown in fig. 19.

To explore the emergent component of the program, variations in the order of rule selection and application were experimented with. By transforming, combining, and replacing layout shapes and rules, the generated designs can be made more interesting. As an example, various alternative designs that rely on the same principles of Mamluk madrasa design could be generated for al-Ashraf Barsbay Madrasa. One of these is shown in fig. 20, which shows the plan of the madrasa after changing the location of the main entrance, the *derka* and the tomb, and after varying the exterior layout shape.

Applying the automated grammar program highlights and explains through repetition the numerical, relational, and morphological aspects of plan organization of Mamluk architecture, which are usually neglected in descriptions of the style. Mamluk madrasas built in Egypt, Syria and Palestine have many morphological similarities. As shown by the derived example, the location of some components, such as the courtyard and the main Iwans, represents a major factor in shaping the grammar of all architectural compositions of Mamluk madrasa architecture. In addition, visual principles such as proportion, symmetry, axis location, and rotation angles are shared by the case studies and are emphasized during the generation process.

Knowledge of the geometric characteristics of the rudimentary vocabulary elements can be gained and enhanced by manipulating the exterior layout and interior space shapes that are offered by the menus for schematic layout and the architectural components.

Knowledge of the scale, proportion, orientation, symmetry, rhythm, and other visual principles can be enhanced by the numerical assignment of the parameters that are required during the rule application process. The shape grammar of Mamluk madrasas is influenced by many visual principles. These include symmetry, unit-to-whole ratio, repetition, addition and subtraction, geometry and grid structure, hierarchy, complexity, and orientation of parts. The major principles are briefly described as follows:

Asymmetric balance: As shown by the example derived, the typical symmetry that is associated with the sacred precedents is substituted by a concept of order that provides balance between parts which form a coherent whole. This concept was achieved through the use of multiple axes of various directions to control all topological relationships of the vocabulary of Mamluk madrasa components. Balance is achieved by geometry when a dominant form, such as the dome on the mausoleum, balances a massive cuboid that is dominant by its size. Although asymmetry is the general characteristic, Mamluk architecture exhibit some levels of symmetry on the organization of the internal layout. The precedents studied reveal central symmetry in which major axes meet in a focal point that is typically located at the centre of the courtyard and controls the layout of the Iwans. However, symmetry in Mamluk madrasas was not perfect. It was applied to control the organization of some internal components of the floor plan rather than the overall exterior shape.

Rhythm: Repetition, expressed as a rhythm of component distribution, can be found in many instances of Mamluk madrasa floor plans. It is expressed in the form of addition or division of a whole, or simply represented as a series of ratios. Elements of the plan were repeated by using the ratios of 1:1; or less frequently 1:2.57, 1:1.6 (approximately the golden section), or 1:3.14 (1: π).

Orientation of elements: A major factor of coherence in Mamluk madrasa architecture depends on grouping elements which have the same function on the same orientation axis. The major axis of orientation is dictated by the angle of Qibla direction. Aagain, this is due to the connection between education and religion in Mamluk madrasa architecture.

Hierarchy: Hierarchy was achieved in Mamluk madrasas by ordering primary components (such as the main courtyard) in the most central position and the secondary elements (such as cells around courtyard) in the less dominant locations.

Simplicity vs. Complexity: Although all interior spaces are simply shaped, the overall layout has a complex form. This complexity refers to two features of the layout which prevent the fragmentation of the secondary components:

1. The change of angle was used as a planning technique to create a new axis of orientation to organize the minor vocabulary elements. While the primary components were directed to the Qibla direction, the secondary ones were made more dynamic by changing their directions to distinguish them visually.

2. The large central space acts as a reference point or datum line for all other plan components.

Understanding the morphology and the design process of individual Mamluk buildings may enhance the explanation of town and city growth in Islamic cultures. By projecting component structure and relations in a Mamluk building on various buildings and their interrelationships in town planning, new layers of interpretation can be added. For example, the previous characteristics such as space hierarchy and angle manipulation can be observed as planning trechniques in some Islamic cities.

The example derived in this section illustrates the time-saving advantage of the program. For example, the manual generation of the floor plan required eight hours as opposed to one hour required for the computer-aided derivation of the same plan. The computer implementation presented in this paper represents a tool for designing an accurate and multi-layered madrasa floor plan.

Using the program conveys a message to the user that the process is as important as, if not more important than, the final product. It facilitates the exploration of the design process and product morphology analysis to better understand the beauty underlying its design. The developed grammar with its systemized components (vocabularies, mathematical and topological relations, rules, and initial shapes) enhances knowledge of *design science* through an algorithm-based framework that performs instantiation, transformation, and combination, as well as set operations of shapes and through connecting style attributes with principles of visual composition.

Although the generative system has good predictive, derivative, and descriptive powers, it has limitations when it comes to explaining the historical, social, and other symbolic and semantic considerations in the architectural composition. A future extension of the program may link the historical and social issues to the syntactic grammar developed in this paper in the form of attributed rules that augment semantic interpretations of the grammatical formalisms. Such augmentation can take the form of conditional or context-sensitive rule application. As such, restrictions can be added to guide the users about which rules to select at each step and in what context to apply them.

6 Conclusion

It has been shown in this paper that a Rule-Based Design (RBD) system can be constructed from the information inferred from case studies to guide the design process. Most of the design information embedded in precedent designs is not in a format that is viable for direct reuse. This paper has suggested a five-phase system of rules to enable restructuring and representing of the unstructured information embedded in precedents in the form of reproductive and recursive rules. Such a representation helps in the regeneration of existing precedents and in the generation of new designs that belong to the same stylistic prototypes. The system makes it possible to experiment with various combinations of rule application to explore new and unexpected alternatives.

After the theoretical formulation of the precedent-based generative system, it is translated into a computer-aided program. The program facilitates the automatic exploration of the aesthetic values of existing Mamluk precedents, as well as the generation of emergent examples of the same style.

This paper claims that a large portion of the rules used to define a style are based on the logic of form-making. The system developed supports this claim by articulating two types of form-making rules. These include the planning-based rules and the aesthetic-related parameters and constraints. The first type is concerned with the procedural aspects and the compositional principles and is expressed as a set of organizational axes and focal points of form-making and is translated into direct rules. The second focuses on the numerical assignment of the dynamic parameters.

The research has shown that an integrated system that is based on a combination of a top-down approach of extracting information from precedents and a bottom-up approach of representing the extracted data as a set of rules can become a powerful tool for form analysis and derivation. The indeterministic nature of the system, coupled with its parametric structure, makes it flexible and dynamic enough to come up with unexpected emergent designs. The theoretical framework as well as the computerized program can be added to the design toolbox to communicate form structure, design process, principles of aesthetic composition and stylistic alphabet, and syntax of architectural design. Limitations of the system include the lack of representation of the social, historical, symbolic, and semantic issues in the form-making process.

A future extension of this research may focus on studying sets of precedents of different styles and developing a multi-style automated generative system and style editor that compares, evaluates, analyzes, and produces architectural designs of various styles. Another extension might take into consideration the morphology of façades or the three-dimensional spatial organizations of the Mamluk precedents as well as their impacts on plan composition. A comparative study could also be conducted to find commonalities or correspondences between the grammar of Mamluk madrasas and that of other madrasas (such as Western madrasas). Furthermore, connections between this research and Problem-Based Learning or E-Learning settings could be emphasized in future research.

Notes

1. According to structuralism [Caws 1988], objects exist in groups that collectively exhibit commonalities in their attributes. Such objects can be grouped into systems which can be defined recursively. Consequently, to understand systems it is necessary to investigate the internal as well as the external relationships of the system components and the structures that underlie their grouping.

References

AGARWAL, M., and J. CAGAN. 1998. A blend of different tastes: the language of coffee makers. *Environment and Planning 'B': Planning and Design* **25**: 205-226.

AL-JOKHADAR, A. 2004. Shape Grammar: An Analytical Study of Architectural Composition Using Algorithms and Computer Formalisms (The Morphology of Educational Buildings in Mamluk Architecture). Unpublished Master Thesis, Department of Architecture, Jordan University of Science and Technology, Irbid, Jordan.

CAWS, P. 1988. *Structuralism: The Art of the Intelligible.* Atlantic Highlands: Humanities Press International.

CHASE, S.C. 1989. Shapes and shape grammars: from mathematical model to computer implementation. *Environment and Planning 'B': Planning and Design* **16**: 215-242.

DOCHY, F., M. SEGERS, P. VAN DEN BOSSCHE and K. STRUYVEN. 2005. Students' Perceptions of a Problem-Based Learning, Environment, *Learning Environments Research* **8**: 41–66.

DUARTE, J. P. 2005. Towards the mass customization of housing: the grammar of Siza's houses at Malagueira. *Environment and Planning 'B': Planning and Design* **32**(3): 347 – 380.

EILOUTI, B. 2001. Towards a Form Processor: A Framework for Architectural Form Derivation and Analysis Using a Formal Language Analogy. Ph.D. Dissertation, University of Michigan.

EILOUTI, B. and A. AL-JOKHADAR. 2007. A Generative System for Mamluk Madrasa Form-Making. *Nexus Network Journal* **9**, 1: 7-30.

FLEMMING, U. 1987. More than the sum of its parts: the grammar of Queen Anne houses *Environment and Planning B: Planning and Design* **14**: 323-350.

GIPS, J. 1975. *Shape Grammars and their Uses*. Basel: Birkhäuser.

———. 2000. *Computer Implementation of Shape Grammars*. Boston: Computer Science Department, Boston College.

GLASSIE, H. 1975. *A Structural Analysis of Historical Architecture*. Knoxville: University of Tennessee Press.

The Islamic Methodology for the Architectural and Urban Design. 1991. Proceedings of the 4th seminar, Rabat, Morocco. Organization of Islamic Capitals and Cities.

KONING, H. and J. EISENBERG. 1981. The language of the prairie: Frank Lloyd Wright's prairie houses. *Environment and Planning: B* **8**: 295-323.

KRISHNAMURTI, R. 1980. The arithmetic of shapes. *Environment and Planning 'B': Planning and Design* **7**: 463-484.

———. 1981a. The construction of shapes, *Environment and Planning 'B': Planning and Design* **8**: 5-40.

———. 1981b. *SGI: A Shape Grammar Interpreter*. Research Report, Centre for Configurational Studies, The Open University, Milton Keynes, UK.

———. 1992a. The maximal representation of a shape. *Environment and Planning 'B': Planning and Design* **19**: 267-288.

———. 1992b. The arithmetic of maximal planes. *Environment and Planning 'B': Planning and Design* **19**: 431-464.

KRISHNAMURTI, R. and C. GIRAUD. 1986. Towards a shape editor: the implementation of a shape generation system. *Environment and Planning 'B': Planning and Design* **13**: 391-403.

OSMAN, M.S. 1998. Shape grammars: simplicity to complexity. Paper presented in University of East London, London.

http://ceca.uel.ac.uk/cad/student_work/msc/ian/shape.html)

PIAZZALUNGA, U. and P.I. FITZHORN. 1998. Note on a three–dimensional shape grammar interpreter, *Environment and Planning B: Planning and Design* **25**: 11–33.

STINY, G. 1975. *Pictorial and Formal Aspects of Shape and Shape Grammars*. Basel: Birkhäuser.

———. 1980. Introduction to shape and shape grammars, *Environment and Planning 'B': Planning and Design* **7**: 343-351.

———. 1994. Shape rules: closure, continuity and emergence. *Environment and Planning 'B': Planning and Design*, vol. 21, Pion Publication, Great Britain, pp. s 49- s 78, 1994

TAPIA, M. A. 1996. From Shape to Style. Shape Grammars: Issues in Representation and Computation, Presentation and Selection. Ph.D. Dissertation, Department of Computer Science, University of Toronto, Toronto.

———. 1999. A visual implementation of a shape grammar system. *Environment and Planning 'B': Planning and Design* **26**: 59-73.

TZONIS, A. and I. WHITE, eds. 1994. *Automation Based Creative Design: Research and Perspectives*, London: Elsevier Science.

About the authors

Buthayna H. Eilouti is Assistant Professor and Assistant Dean at the Faculty of Engineering in Jordan University of Science and Technology. She earned a Ph.D., M.Sc. and M.Arch. degrees in Architecture from the University of Michigan, Ann Arbor, USA. Her research interests include Computer Applications in Architecture, Design Mathematics and Computing, Design Pedagogy, Visual Studies, Shape Grammar, Information Visualization, and Islamic Architecture.

Amer Al-Jokhadar is a Part-Time Lecturer at the College of Architecture and Design at the German-Jordanian University in Amman, and an architect in TURATH: Architectural Design Office. He earned M.Sc. and B.Sc. degrees in Architectural Engineering from Jordan University of Science and Technology. His research interests include Shape Grammar, Mathematics of Architecture, Computer Applications in Architecture, Islamic Architecture, Heritage Conservation and Management, Design Methods.

Dirk Huylebrouck

Department for Architecture
Sint-Lucas
Paleizenstraat 65-67
1030 Brussels BELGIUM
Huylebrouck@gmail.com

Keywords: curve fitting,
design analysis, Antoni Gaudí,
linear algebra, generalised
inverses, least squares,
hyperboloid, catenary, parabola,
golden number

Research

Curve Fitting in Architecture

Abstract. It used to be popular to draw geometric figures on images of paintings or buildings, and to propose them as an "analysis" of the observed work, but the tradition lost some credit due to exaggerated (golden section) interpretations. So, how sure can an art or mathematics teacher be when he wants to propose the profile of a nuclear plant as an example of a hyperboloid, or proportions in paintings as an illustration of the presence of number series? Or, if Gaudi intended to show chain curves in his work, can the naked eye actually notice the difference between these curves and parabolas? The present paper suggests applying the "least squares method", developed in celestial mechanics and since applied in various fields, to art and architecture, especially since modern software makes computational difficulties nonexistent. Some prefer jumping immediately to modern computer machinery for visual recognition, but such mathematical overkill may turn artistic minds further away from the (beloved!) tradition of geometric interpretations.

Introduction

Until about twenty years ago, it was common to draw all kinds of geometric figures on images of artworks and buildings. Usually, simple triangles, rectangles, pentagons, or circles sufficed, but sometimes more general mathematical figures were used, especially after fractals became trendy. Recognizing well-known curves and polygons was seen as a part of the "interpretation" of an architectural edifice or painting. Eventually, segments of lines occurred in certain proportions, among which the golden section surely was the most (in)famous. Diehards continue this tradition, though curve drawing has lost some credit in recent times, in particular due to some exaggerated interpretations of the golden section.

Today, journals dealing with the nexus of mathematics and architecture tend to reject these "geometric readings in architecture". Their point of view may be justified by several shortcomings of common research in the field:

i. An architect may have had the intention of constructing a certain curve or surface, and even discuss these intentions in his plans, but for structural, technical or various practical reasons, the final realization may not conform to that intention.

Example: a nuclear plant is said to have the shape of a hyperboloid, but engineers modify its top to reduce wind resistance. So, how close to a hyperbola is the final silhouette of the structure?

ii. An artist may have used a certain proportion, consciously or unconsciously, so that when such a "hidden" proportion or curve is discovered, even the author of the artwork may dispute its use.

1590-5896/07/010059-12 DOI 10.1007/s00004-006-0029-3

Example: It happens that rational subdivisions are discovered in impressionist paintings. Some artists claim they intended only pure emotion, disliking art-school-like subdivisions such as the golden section. However, some critics can still demonstrate the presence (or absence) of this proportion using arguments similar to widely accepted interpretations of other artwork.

    iii.   In the absence of accurate written sources about a work of art or an architectural design, how does one decide whether an interpretation is correct or not? Is it merely a matter of a subjective interpretation, a posteriori?

Example: in a discussion about pointed arches, how does one decide if an arch was pointed or not, without being mislead by cultural penchants?

Despite this unenthusiastic introduction, this author still believes there could be positive reasons for continuing the study of geometric views on art and architecture:

    i.   An engineer may want to reveal structural properties of the building.

Example: the difference between a parabola or chain curve when a cable supports a horizontal bridge, or only its own weight.

    ii.   A mathematician may want to express what he finds beautiful in his preferred mathematical language and there is no reason why only literary minds should enjoy this privilege.

Example: Horta was inspired by spirals he founds in plants. These spirals have a precise mathematical description, and though it is not certain that Horta intended to represent these curves, the mathematical interpretation may show a relationship with other sources of inspiration by other architects. If another architect was inspired by a nautilus shell, then both, in fact, had a similar inspiration, at least, as far as the mathematician is concerned.

    iii.   An artist or architect may have been inspired by a curve or a proportion he finds elegant.

Example: well-defined proportions may have been used to subdivide a canvas or to construct buildings. This can help to collect information on how the work was done, and, for instance, to determine the unit length used (non-standard foot, cubit).

In the present paper, a similarity is proposed between these geometric studies in architecture and the history of (celestial) mechanics. It is suggested that the so-called "least squares method" developed in that field could be applied to examples in art as well. Of course, it can be argued that, unlike celestial mechanics with its involved applications, such a mathematical method would represent serious overkill with respect to the intended straightforward artistic applications. However, if this was true in the past, modern software allows slimming down computational aspects to some simple computer clicks.

Campbell and Meyer's account on curve fitting

A well-known example in the history of curve-fitting is the story of how Carl Friedrich Gauss found a "lost planet". Campbell and Meyer [1979] gave the following excellent account illustrating curve fitting, in the context of celestial mechanics and linear algebra.

In January 1801, astronomer G. Piazzi briefly observed a "new planet", Ceres, and astronomers tried, in vain, to relocate it for the rest of 1801. Only Gauss could correctly predict when and where to look for the lost planet but because he waited until 1809 to publish his theory, some accused him of sorcery.

For the sake of exposition, Campbell and Meyer simplified the problem to an elliptical orbit of which four observations were made: (-1,1), (-1,2), (0,2), (1,1). In this example, obviously, there must have been some errors, of whatever kind (inaccurate observations or instruments, or mere bad luck), since two data correspond to -1. Still, the question now is to find the ellipse

$$\frac{x^2}{a^2} + \frac{y^2}{b^2} = 1 \ ,$$

approximating the four data as closely as possible.

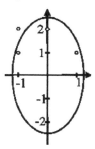

Fig. 1. Campbell and Meyer's example of finding the ellipse closest to four data points

Putting $b_1 = 1/a^2$ and $b_2 = 1/b^2$, implies the four data points (x_i, y_i), $i = 1, ...4$, should make the error $x_i^2 b_1 + y_i^2 b_2 - 1$ minimal. In matrix notation, it means $\mathbf{X.b\text{-}j}$ should be minimal, where:

$$\mathbf{X} = \begin{bmatrix} x_1^2 & y_1^2 \\ x_2^2 & y_2^2 \\ x_3^2 & y_3^2 \\ x_4^2 & y_4^2 \end{bmatrix}, \mathbf{b} = \begin{bmatrix} b_1 \\ b_2 \end{bmatrix}, \mathbf{j} = \begin{bmatrix} 1 \\ 1 \\ 1 \\ 1 \end{bmatrix}.$$

In the general case, it turns out $\mathbf{Xb\text{-}j}$ admits a unique least squares solution of minimal norm $\mathbf{b} = \mathbf{X}^+\mathbf{j}$, while the precision of fit can be measured by comparing the image, $\mathbf{Xb}$, to the desired value $\mathbf{j}$. Thus, the fraction $R^2 = \|\mathbf{Xb}\|^2/\|\mathbf{j}\|^2$ is used as a quantity giving an idea about the accuracy of the proposed approximation. The notation $\mathbf{X}^+$ represents the so-called "generalized inverse" in the sense of Moore-Penrose, which corresponds to the regular inverse if it exists (that is, when $\mathbf{X}$ is invertible, and thus, when an exact solution can be computed). We will not bother here about explaining computational aspects of this $\mathbf{X}^+$ when $\mathbf{X}$ is not invertible, since mathematical software such as MATHEMATICA^{TM}, allows computing this generalization of the notion of the inverse, called "pseudo inverse", without effort.

In the given example, $\mathbf{X} = \begin{bmatrix} 1 & 1 \\ 1 & 4 \\ 0 & 4 \\ 1 & 1 \end{bmatrix}$ and $\mathbf{b} = \mathbf{X}^+\mathbf{j} = \begin{bmatrix} 7 \\ \frac{11}{2} \\ \frac{2}{11} \end{bmatrix}$ and $\mathbf{R}^2 \approx 0.932$.

The set-up can be modified to find, for instance, the nth degree polynomial

$$y = b_0 + b_1 x + b_2 x^2 + b_3 x^3 + \ldots + b_n x^n$$

which best fits a set of given points (x_i, y_i). In this case, the expression $\mathbf{X.b\text{-}j}$ should be minimal, and the involved matrices are:

$$\mathbf{X} = \begin{bmatrix} 1 & x_1 & x_1^{\ 2} & \ldots & x_1^{\ n} \\ 1 & x_2 & x_2^{\ 2} & \ldots & x_2^{\ n} \\ \vdots & \vdots & \vdots & \ddots & \vdots \\ 1 & x_m & x_m^{\ 2} & \ldots & x_m^{\ n} \end{bmatrix}, \mathbf{b} = \begin{bmatrix} b_1 \\ b_2 \\ \vdots \\ b_n \end{bmatrix}, \mathbf{j} = \begin{bmatrix} 1 \\ 1 \\ \vdots \\ 1 \end{bmatrix}.$$

Again, $\mathbf{X.b\text{-}j}$ admits a unique "least squares solution of minimal norm" $\mathbf{b} = \mathbf{X}^+\mathbf{j}$ while the precision of fit can be measured by $\mathbf{R}^2 = \|\mathbf{Xb}\|^2 / \|\mathbf{j}\|^2$.

Application 1: a chain curve approximates Gaudi's Paelle Guell better than a parabola

Gaudi is known for his use of hyperbolic cosine functions. He used bags suspended by ropes which he inverted to get chain curves. However, the question remains whether or not a non-informed spectator can actually see this in Gaudi's work, that is, if a spectator can tell that a Gaudi gate, entrance or window (as for example in the Paelle Guell, fig. 2a) has a hyperbolic cosine shape. Couldn't he conclude that the observed curves are, for instance, parabolas?

Fig. 2a

I proposed this question during a course at the Sint-Lucas Institute for Architecture (Belgium), and student Sil Goossens came up with the following example, for which he concluded the parabola with equation y = -0.811x² + 5.704x + 2.643 came pretty close, but not close enough. He claimed "the Gaudi entrance should be considered as a chain curve".

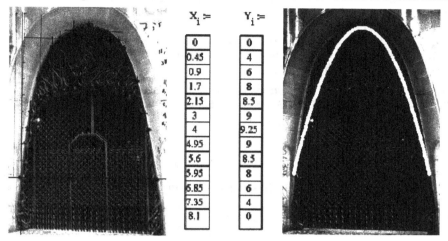

$X_i :=$	$Y_i :=$
0	0
0.45	4
0.9	6
1.7	8
2.15	8.5
3	9
4	9.25
4.95	9
5.6	8.5
5.95	8
6.85	6
7.35	4
8.1	0

Fig. 2b. Gaudi's Paelle Guell gate in Barcelona, some measurements made by a student, and a parabola fitting well, but not well enough

How can this idea be substantiated? First, checking the coordinates, they look fairly correct, but we make a change of coordinates to situate the top in (0, 1), since this is a more straightforward choice when a hyperbolic cosine is expected. In addition, the x-axis is scaled down by 10% so that all x<1, as this can make the use of series easier (though it will later turn out to be useless). A MATHEMATICA™ input provided the following :

In:

```
x1=-.4; x2=-.36; x3=-.31; x4=-.235; x5=-.185; x6=.105; x7=0; x8=.09; x9=.16;
    x10=.2; x11=.29; x12=.34; x13=.4;
j={-8, -4, -2, -.25, .25, .7, 1, .7, .15, -.4, -2.55, -4.45, -8};
X={{1,x1,x1^2},{1,x2,x2^2},{1,x3,x3^2},{1,x4,x4^2},{1,x5,x5^2},{1,x6,x6^2},{1,x7,x
    7^2},{1,x8,x8^2},{1,x9,x9^2},{1,x10,x10^2},{1,x11,x11^2},{1,x12,x12^2},{1,x13,x
    13^2}};
B=PseudoInverse[X].j
Norm[X.B]^2/Norm[j]^2
```

Out:

```
    {1.74011,-1.24714,-53.9541}
    0.96035
```

that is, a parabola with equation y = 1.74 - 0.125x -54x². It fits well, since R² = 96%. We can try to fit a hyperbolic cosine, using:

```
X={{1,Cosh[x1] }, …,{1,Cosh[x13]}};
B=PseudoInverse[X].j
Norm[X.B]^2/ Norm[j]^2
```

Now: y = 107.5-105.8·Cosh[x], and it fits at 95.3%. Not only do we wonder if this 1% of difference could be noticed, but further, the result seems to contradict the student's conclusion. Looking carefully at the picture, we see the photo is slightly skew with respect to the observer, and because of this lack of symmetry, the proposed hyperbolic cosine will never give a nice fit, of course. Thus, we can try to take into account the imperfections in the picture, and propose a symmetric interpretation, estimating the second coordinates by modifying the given values slightly:

x13=-x1; x12=-x2; x11=-x3; x10=-x4; x9=-x5; x8=-x6;
j={-7.5, -4.4, -2.55, -.4,.25, .75, 1, .75, .25, -.4, -2.55, -4.4, -7.5};
X={ {1,x1,x1^2,x1^3,x1^4,x1^5,x1^6},..., {...,x13^6}};

This produced the following outcome:

$$y = 1.019 - 6.115 \cdot 10^{-16} x - 20.95 x^2 - 1.10 \cdot 10^{-13} x^3 - 60.7 x^4 - 9.9 \cdot 10^{-13} x^5 - 8.66.96 x^6$$

fitting at 99.875%. If we keep in mind the series for Cosh(x):

$$2 - \text{Cosh}(x) = 2 - \left(1 + \frac{x^2}{2!} + \frac{x^4}{4!} + \frac{x^6}{6!} \cdots \right) = 1 - \frac{x^2}{2!} - \frac{x^4}{4!} - \frac{x^6}{6!} \cdots$$

we see that:

$$y \approx 1 - 21 x^2 - 61 x^4 - 867 x^6 \approx 1 - 21.2! \frac{x^2}{2!} - 1454 \frac{x^4}{4!} - 867.6! \frac{x^6}{6!} \approx 1 - \frac{(6.48x)^2}{2!} - \frac{(6.19x)^4}{4!} - \frac{(9.3x)^6}{6!}$$

Thus, we should try a different hyperbolic cosine, using a Cosh[a.x] expression, where the coefficient a is somewhere in the 6.19 or 9.3 range. Now y = 1.34 - 0.36·Cosh[9.7x] fits at 99.88% and this indeed beats the closest possible parabola, which is y = 1.84 - 52.12x^2, fitting at only 96.75%. This 3% difference in closeness of fit can indeed be noticed, as the illustration shows.

Fig. 3. The catenary and parabola compared to the initial picture. Data were changed putting the picture in a frontal position: only the left half is to be considered.

Application 2: Gaudi's Collegio Teresiano can be seen either as a catenary or as a parabola

The previous result was nevertheless contradicted by another architecture student, Klaas Vandenberghe, who claimed some of Gaudi's work shows no hyperbolic cosines, but parabolas. His examples were the parallel arcs of the Collegio Teresiano. Again, the teacher was asked to set things straight and to decide who was right.

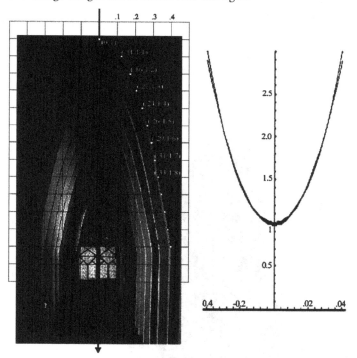

Fig. 4: The original picture and the obtained catenary and parabola, which are so close they cannot be distinguished

The procedure is similar: we measure the data as well as possible on the provided picture, and use the pseudo-inverse algorithm. We start with a 6th degree polynomial:

$$y = 1.026 - 3.33 \cdot 10^{-15} x + 5.3x^2 + 5.3 \cdot 10^{-14} x^3 + 29.5x^4 + 1.81 \cdot 10^{-13} x^5 - 100.9x^6,$$

for which the fit is R^2 = 99.99%. Thus:

$$y \approx 1 + 5.3x^2 + 29.5x^4 + 100.9x^6 \approx 1 + 5.3 \cdot 2! \frac{x^2}{2!} + 29.5 \cdot 4! \frac{x^4}{4!} + 100.9 \cdot 6! \frac{x^6}{6!} \dots$$

$$\approx 1 + \frac{(3.2x)^2}{2!} + \frac{(5.1x)^4}{4!} + \frac{(6.4x)^6}{6!} \dots$$

This again reminds us of the hyperbolic cosine series, and

$$y = -0.7468 + 1.75.\mathrm{Cosh}[2.8x]$$

gives a 99.988% fit. A parabola can fit nicely too, as $y = 0.985 + 7.63x^2$, shows, with a 99.985% fit. This difference of 0.003% cannot be distinguished with the naked eye.

Application 3: The profile of a nuclear power plant is a hyperbola, or even, an ellipse

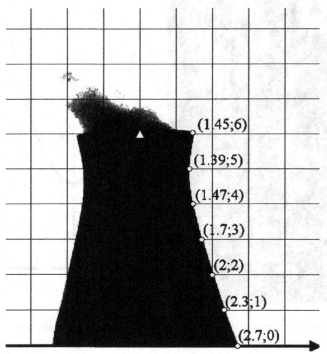

Fig. 5. Determining the profile of a nuclear plant

Consider fig. 5, from which the following values were drawn:

$$x_1 = 2.7; x_2 = 2.3; x_3 = 2; x_4 = 1.7; x_5 = 1.47; x_6 = 1{,}39; x_7 = 1.45;$$
$$y_1 = 0; y_2 = 1; y_3 = 2; y_4 = 3; y_5 = 4; y_6 = 5; y_7 = 6.$$

We compute the general conic section with equation

$$x^2 b_1 + y^2 b_2 + x b_3 + y b_4 + xy b_5 - 1 = 0$$

The minimal norm least squares solution follows from:

X={{x1^2,y1^2,x1,y1,x1*y1},{x2^2,y2^2,x2,y2,x2*y2},{x3^2,y3^2,x3,y3,x3*y3},{x4
^2,y4^2,x4,y4,x4*y4},{x5^2,y5^2,x5,y5,x5*y5},{x6^2,y6^2,x6,y6,x6*y6},{x7^2,y7
^2,x7,y7,x7*y7}};
j={1,1,1,1,1,1,1};
B=PseudoInverse[X].j
Norm[XB]^2/Norm[j]^2

It turns out to be

$$-0.11x^2 - 0.0105y^2 + 0.67x + 0.19y - 0.06xy = 1 \text{ at } 99.9996\%.$$

Surprisingly, this is an ellipse. Still, if we lift the x-axis over 5 units, we can propose the standard form equation for a hyperbola: $x^2b_1 + y^2b_2 = 1$, where b_2 should be negative. Now:

In:

x1=2.7; x2=2.3; x3=2; x4=1.7; x5=1.47; x6=1.39; x7=1.45;
y1=-5; y2=-4; y3=-3; y4=-2; y5=-1; y6=0; y7=1;
X={{x1^2,y1^2},{x2^2,y2^2},{x3^2,y3^2},{x4^2,y4^2},{x5^2,y5^2},{x6^2,y6^2},{x
7^2,y7^2}};
j={1,1,1,1,1,1,1};
PseudoInverse[X].j
Norm[X.PseudoInverse[X].j]^2/Norm[j]^2

Out

{0.508687, -0.108291}
0.998771

This is indeed a hyperbola, which comes as close as 99.8%. Thus, we can faithfully claim that the shape of a nuclear plant is a hyperbola, and that the adaptations on the top of the building, for reasons of resistance to wind, can hardly be detected.

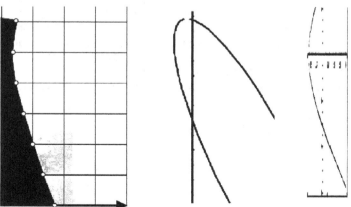

Fig. 6. The profile of a nuclear plant, and the approximating shapes, as an ellipse (middle) and a hyperbola (right)

Application 4: golden rectangles

Golden section "interpretations" of the Mona Lisa are popular and the influence of *The da Vinci Code* surely did not diminish its appeal. Even Wasler's excellent book [2001] presents a strange inexplicable shift in the rectangles covering the body, so we'll concentrate on the facial proportions, which seem more observable.

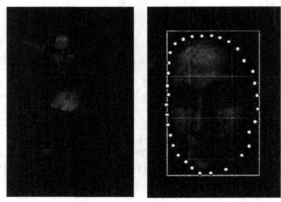

Fig. 7. The cover of Walser's book (left), and a set of data points for the Mona Lisa (right)

Pointing the cursor on the scanned image, we get numbers that have rather useless and far too many decimals places, but it takes more time and energy to judge about their utility than to copy them from the computer output:

x1 = 17.496; y1 = 4.871; x2 = 19.129; y2 = 7.143; x3 = 19.200; y3 = 10.551;
x4 = 20.549; y4 = 13.107; x5 = 21.259; y5 = 16.160; x6 = 21.046; y6 = 19.426;
x7 = 20.549; y7 = 22.337; x8 = 20.052; y8 = 25.248; x9 = 18.064; y9 = 27.804;
x10 = 16.573; y10 = 30.644; x11 = 14.372; y11 = 32.490; x12 = 12.313; y12 = 33.413;
x13 = 10.396; y13 = 33.910; x14 = 8.195; y14 = 33.839; x15 = 5.852; y15 = 33.697;
x16 = 3.722; y16 = 33.058; x17 = 1.947; y17 = 31.993; x18 = .740; y18 = 29.792;
x19 = 0.456; y19 = 27.378; x20 = .243; y20 = 24.822; x21 = 0.243; y21 = 12.397;
x22 = 0.811; y22 = 9.912; x23 = 1.450; y23 = 7.427; x24 = 2.657; y24 = 5.013;
x25 = 4.361; y25 = 3.380; x26 = 5.710; y26 = 1.889; x27 = 7.688; y27 = 0.571;
x28 = 10.224; y28 = 0.571; x29 = 13.805; y29 = 1.541;
X = {{x1^2, y1^2, x1, y1, x1*y1}, ... {x29^2, y29^2, x29, y29, x29*y29}};
j = {1, ... 1};
PseudoInverse[X].j
Norm[X.PseudoInverse[X].j^2/Norm[j])^2

The result is the ellipse

$$- 0.0078x^2 - 0.0033y^2 + 0.1686x + 0.1241y - 0.0009xy = 1$$

fitting at R^2 = 99.55%.

Now the minimum corresponds to x=10.8697..., while the maximum corresponds to 8.99537..., and the difference in y-values is 34.014. The (horizontal) width is 20.9757...-

(-1.110...) = 22.09, and the proportion is N[34.014/22.09] = 1.54.... This is closer to, say, 1.5, than to the golden number of 1.618...

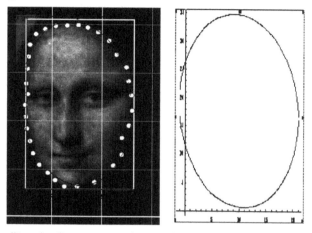

Fig. 8. An ellipse that fits to 99.55%, showing that the vertical to horizontal ratio is 1.54

Conclusion

The least-squares method, known from celestial mechanics and since applied in the most various areas, could well be used for geometric interpretations in art as well, especially since modern software makes the computational difficulties nonexistent. One can only wonder why such a widespread and straightforward mathematical technique has not been applied before. Today, some prefer jumping immediately to modern computer machinery in visual recognition of patterns, but such mathematical overkill is not necessary and can even turn artistic minds away from the (beloved!) topic of geometric interpretations in architecture.

References

CAMPBELL, S.L. and C.D. MEYER, Jr. 1979. *Generalized Inverses of Linear Transformations.* London: Pitman Publishing Limited.

MARKOWSKY, George. 1992. Misconceptions about the Golden Ratio. *The College Mathematics Journal* **23** (January 1992): 2 - 19.

SPINADEL, Vera W. de. 1998. *From the Golden Mean to Chaos.* Buenos Aires: Nueva Libreria S.R.L.

WALSER, Hans. 2001. *The Golden Section.* The Mathematical Association of America.

WASSELL, Stephen R. 2002. The Golden Section in the Nexus Network Journal. *Nexus Network Journal* **4**, 1 (Winter 2002): 5-6.

About the author

Dirk Huylebrouck obtained his Ph.D. in algebra at the University of Ghent. He worked in Africa for about twelve years, at the Universities of Congo (Bukavu, Kinshasa) and Burundi, interrupted by assignments at the Portuguese University of Aveiro and at the Overseas Divisions of American universities in Europe (such as Maryland and Boston University). Since about ten years, he teaches at the Department for Architecture Sint-Lucas Brussels (Belgium). His research on generalized inverses was awarded by a quotation in the book "Current Trends in Matrix Theory 1987", and his study of zèta(3) the "Lester Ford Award 2002" of the Mathematical Association of America. As editor of the "Mathematical Tourist" column in *The Mathematical Intelligencer*, he paid attention to historical and artistic aspects of mathematics from Tibet over Africa and Europe to Siberia. A mathematical artifact from Central Africa inspired him to write a book, from which he presented many talks, publications and TV-collaborations. He currently tours around Belgium in a multicultural mathematics show together with the percussion trio "Dakar Electric".

Giulio Magli

Dipartimento di Matematica
Politecnico di Milano
P.le L. da Vinci 32
20133 Milano, Italy
giulio.magli@polimi.it

Keywords: orthogonal town
planning, polygonal walls

Research

Non-*Orthogonal* Features in the Planning of Four Ancient Towns of Central Italy

Abstract. Several ancient towns of central Italy are characterized by imposing circuits of walls constructed with the so-called polygonal or "cyclopean" megalithic technique. The date of foundation of these cities is highly uncertain; indeed, although they all became Roman colonies in the early Republican centuries (between the fifth and third centuries B.C.) their first occupation predates the Roman conquest. It is the aim of the present paper to show – using four case-studies – that these towns still show clear traces of an archaic, probably pre-Roman urbanistic design, which was *not* based on the orthogonal "rule", i.e., the town-planning rule followed by the Greeks, Etruscans and Romans. Rather, the layouts appear to have been originally planned on the basis of a *triangular*, or even star-like, geometry, which therefore has a center of symmetry and leads to radial, rather than orthogonal, organization of the urban space. Interestingly enough, hints – so far unexplained – pointing to this kind of town planning are present in the works by ancient writers as important as Plato and Aristophanes, as well as in the comment to the *Æneid* by Marius Servius.

1 Introduction

At the end of the Hellenic Dark Age (around the eighth to seventh centuries B.C.) the Greeks began the expansion which soon led to the foundation of several towns ("colonies") in a wide area of the Mediterranean Sea. *All* such towns have been planned on the basis of an orthogonal grid, which divides the urban space in equal rectangular blocks. This is true, for instance, for the oldest colonies (e.g., Selinunte) already in the early sixth century B.C., while in the fifth century, with the reconstruction after the Persian War (e.g., of Miletus, fig. 1) the orthogonal grid became a rule, theorized, at least according to what has been referred by Aristotle, by the architect Hippodamus [Castagnoli 1971].

The organization of the streets on the basis of orthogonal sectors was developed, more or less simultaneously with the Greeks, by the Etruscans. We can be certain of this because, although most of the Etruscan towns were reorganized, or even completely rebuilt by the Romans (so that their original urbanistic design is uncertain), one Etruscan town was destroyed by the Celts before the Roman expansion: Misa (today Marzabotto). The archaeological excavations at Misa have shown that the town was planned on the basis of an orthogonal grid oriented to the cardinal points within 2° of error [Mansuelli 1965]. Many of the blocks of the grid were traced on the ground but never edified, showing that the planners were foreseeing a wide development of the town (fig. 2).

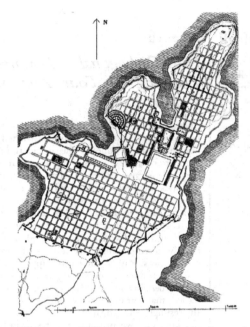

Fig. 1. The layout of Miletus, based on a rigid orthogonal grid

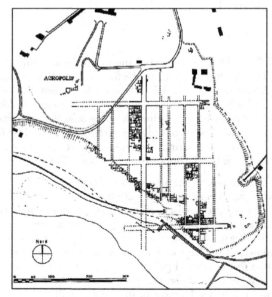

Fig. 2. Plan of Misa (Marzabotto)

It should, of course, be noted that the Etruscans were in close trading contact with the Greeks, so that the degree of the cultural interchange in the process which led to orthogonal town planning is unclear. In any case, Misa shows, *in addition* to the orthogonal grid, the orientation of the streets system to the cardinal points, a thing which is barely visible in Greece. This is a reflection of the complex foundation ritual (to be discussed in section 2) which, at least according to many Roman historians, was elaborated by the Etruscans and directly inherited by the Romans, who combined the orthogonal grid with the inspiring principle of the so-called *castrum* (military camp) criss-crossed by two main roads. The structure of the Roman grid was thus based on two main orthogonal axes, the *Cardus*, oriented (at least in principle) north-south, and the *Decumanus*, oriented east-west, corresponding to four main gates at their ends [Rykwert 1999]. Layouts based on this principle can be seen in *all* the towns of Roman foundation from the middle of the third century B.C. onward (see for instance the plan of Augusta Praetoria, today Aosta, founded around 25 B.C., fig. 3).

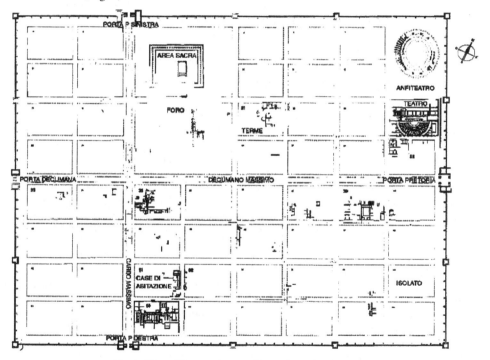

Fig. 3. Plan of Roman Augusta Praetoria, today Aosta

Sometimes the disposition of the grid followed health criteria, such as those recommended by Vitruvius in function of the winds; in some other cases it was rather dictated by symbolic reasons, as shown by the astronomical alignments observed in towns such as Augusta Bagiennorum [Barale et al. 2002] and Bologna [Incerti 1999].

It is worth mentioning that orthogonal town planning remained the rule up to the beginning of the twentieth century, and it is still today considered the best method of town planning, at least in the cases of heavy car circulation [Southworth and Ben-Joseph 1997]. Throughout the world there have been very few exceptions to this rule in the last 2500 years, a notable one being the radial organization of the human space which was a fundamental characteristic of the Inca, as shown by the radial sectors in which the capital of the Inca empire, Cusco, was symbolically divided (see [Magli 2006a], and complete references therein).

All in all this is, briefly, what can be observed in the layout of Greek and Etruscan towns, as well as in the plan of all those towns whose Roman foundation is certain. However, there exist enigmatic and never explained passages of ancient Greek authors as important as Plato and Aristophanes, as well as a somewhat famous comment to the *Æneid* written by Marius Servius, in which these authors seem to refer to a completely different kind of town planning, which is triangular or even radial; up to the present, no convincing explanation of such passages is available. As far as the present author is aware, however, nobody has ever tried to verify if there actually *are* ancient towns showing the traces of a planning based on a triangular, rather than orthogonal, symmetry. The aim of the present paper is to carry out such an analysis, considering as case-studies four among the most beautifully preserved settlements of central Italy, characterized by imposing circuits of walls constructed with the so-called polygonal or "cyclopean" megalithic technique.

2 The Etruscan-Roman foundation ritual

Before entering into the characteristics of the layouts of these towns, it is worth making a brief discussion of the symbolism associated with the town foundation and planning, at least according to the texts which have survived. Indeed, Roman historians such as Varro, Plutarch and Pliny the Elder report that the foundation of towns was governed by a ritual which was directly inherited from the Etruscans and governed by the rules written in the sacred books called *Disciplina*. The *Disciplina* was the collection of writings of the Etruscan religion, which was thought of as having being revealed to humanity by the gods. These books are long lost, but accounts on them have survived (for instance the work *De Divinatione* by Cicero) so that we know that they were composed of three parts. First of all, the *libri haruspicini*, which dealt with divination (the interpretation of God's will) made by the priests called *Aruspexes* by "reading" the flight of the birds and the livers of sacrificed animals, especially sheep; second, the *libri fulgurales*, on the interpretation of thunders, and finally the *libri rituales*, dedicated to all aspects of life, such as the consecration of temples, the division of the people into tribes, and the foundation of towns. The latter consisted in observing the flight of the birds and in tracing the contour of the town by a plough, steps which everybody will recognize in the worldwide famous legend of the foundation of Rome as well. A fundamental part of all the rituals of the aruspexes was the individuation of the *auguraculum*, a sort of terrestrial image of the heavens (*templum*) in which the gods were "ordered" and "oriented". A key "document" about this complex symbolic structure is the so-called *Piacenza Liver* (fig. 4).

The Piacenza Liver is a first century B.C. bronze model of the liver of a sheep, in 1:1 proportions, found in a field near Piacenza in the nineteenth century. The upper surface presents three protuberances (one of them corresponds to the gallbladder); the external perimeter is divided into sixteen sectors, while the surface shows six sectors disposed in a circle, and eighteen further regions; each sector or region contains the name of an Etruscan deity, with some of them repeated (many of these have been identified with the corresponding Roman deities, such as Jupiter or Mars). The lower surface is divided into two regions, having the name of the Sun and of the Moon respectively [Pallottino 1997].

Fig. 4. The Piacenza Liver

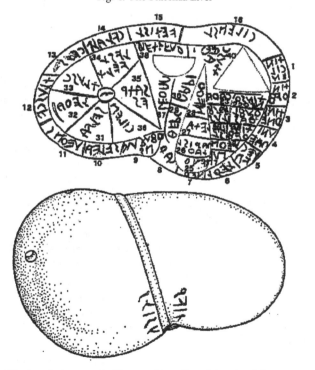

Fig. 5. The inscriptions on the Piacenza Liver (from [Aveni and Romano 1994])

The owner of the Piacenza Liver was certainly an aruspex, and the object was probably used to teach divination, and/or as a help to the memory when doing the exam of the sacrificed animals. It is an extremely important find because its sixteen external divisions show the structure of the Etruscan cosmos (fig. 5).

The deities in each sectors are *ordered* from the most important/benign to those of the underworld; since we know from independent sources that these deities were arranged also in a "geographical" order starting from the north and moving in what we would call the clockwise direction, it follows that the Piacenza Liver was also meant to be "oriented" to the cardinal points [Aveni and Romano 1994, Pallottino 1997]. Once oriented, the liver itself was to become an image of the cosmos reported on the earth, and, by analogy, so was the temple, or the city, that the aruspex was ritually founding. The "center" of the city, called *mundus* by the Romans, was therefore an icon of the center of the world, and contained a "deposit of foundation" in which the first produce of the fields and/or samples of soil from the native place of the founders was buried. Archaeological proofs of the foundation rituals have been discovered in the excavations of the Etruscan towns Misa [Mansuelli 1965] and Tarquinia [Bonghi-Jovino 1998, 2000]. From the Roman period (around the first century B.C.) an example of *auguraculum* is known from the city of Bantia [Torelli 1966]. It is composed by nine stone cylinders (*cippus*) which were disposed on the ground to identify the eight main divisions of the cosmos (a simplified version of the sixteen Etruscan divisions) and the center. The center itself was dedicated to the sun, while the other cylinders carry inscriptions which recall the role of the birds which come from the corresponding direction; for instance, the north-east one says BIVAV that is B[ENE] IU[VANTE] A[VE] (bird bringing a good omen) while the north-west cippus has the inscription CAVAP, that probably means C[ONTRARIA] AV[E] A[UGURIUM] P[ESTIFERUM] (bird who comes from a bad place, bringing a pestiferous omen). Finally, very interesting traces of the foundation ritual, to be discussed in detail below, have been found in the city of Cosa.

The profound relationship between divination, cosmos and foundation ritual is thus indubitable. However, the *radial* division of the Cosmos, reported on the Piacenza Liver, seems to conflict with the *orthogonal* organization of the Etruscan-Roman town space; as mentioned in the introduction, there is also a written text, an enigmatic passage of the Latin writer Servius Marius Honoratus, who describes the town prescribed by the *Disciplina* in a way which it is hard to reconcile with the "squared" town. Servius wrote a important commentary to Virgil's *Æneid* in the fourth century A.D. When commenting on the marvellous verses of the poem in which the hero is urged to admire the construction of the newly founded city of Carthage[1] (in the reality the city was founded by Phoenicians at the end of the ninth century B.C.), Servius writes:

> The experts of the *Etrusca Disciplina* state that those founders of towns who do not plan the layout with three gates, three main streets, and three temples dedicated to Jupiter, Juno and Minerva, cannot be considered as people who obey the rules [author's translation].[2]

Thus, we have here the description of a sort of radial, or at least *triangular* town, with three temples dedicated to the three main gods. The dedication to three gods can be easily explained in the Roman context because Jupiter, Juno and Minerva formed the *Triade Capitolina*, to which usually the main temple was dedicated (although it usually was a *single* temple with *three* cells); however, the town's layout described by Servius, based on the number 3, can hardly correspond to a town planned on an orthogonal grid or, even less, to the Roman Castrum with four main gates and two main streets. As a consequence, this passage has generated much confusion in the scholars who have tried to interpret it. For instance, Bloch [1970] noticed that in Misa (fig. 2) the Acropolis is located in the north, and therefore, one can conceive the idea that, *from* the Acropolis, the quadripartite town would have looked as tripartite, as described by Servius; however, this is quite an *ad hoc* explanation, which holds, if it does at all, only for a specific case. We shall instead see that there are towns in Italy – evidently *more ancient than Misa* – which might have been originally planned in accordance to the rules recalled by Servius.

3 The layouts of four cyclopean towns in central Italy

In a wide area of west-central Italy, which extends from Tuscany to Campania, there exist several towns whose walls were built with the so called *cyclopean* or *polygonal* technique. These walls are constructed of enormous (from one up to twenty tons of weight) stone blocks, cut in polygonal shapes and fitted together without mortar to form a sort of giant puzzle. Such a spectacular technique makes its first appearance during the Bronze Age (around 1500-1300 B.C.) among the Myceneans, and indeed the attribute "cyclopean" comes from Pausania's description of the walls of Mycenae and Tyrins. Although much less famous, the Italian cyclopean towns also achieve the same magnificence and impression of power and pride which characterize the world-famous Mycenaean sites; usually, however, the dimensions of the Italian towns are much *wider* (the perimeter can be as long as 3 km) and the most striking similarity to the Mycenaean buildings is reached in the so-called Acropoli. These are imposing megalithic buildings, situated in a dominant position with respect to the landscape, very similar to the Mycenaean's citadels in dimensions (of the order of some tens of thousands of square metres), form (polygonal, like that of the blocks), accesses (they usually have only one main gate and one postern gate on the opposite side), contents (the interior usually contained a megalithic basement, perhaps a temple or a palace), and in their relationship with the landscape, being visible from very far; last but not least, they are nearly identical in construction technique and show astronomical alignments which can hardly be attributed to the Romans (for more details on the Acropoli of Central Italy see [Magli 2006b, 2006c] and references therein). In spite of all this, almost all polygonal walls in Italy, both in the case of the town walls and in the case of the Acropoli, are currently dated by the archaeologists many centuries after the Myceneans, that is, to the first centuries of the Roman expansion, between the end of the sixth and the third century B.C. (see e.g., [Lugli 1957]). However, with few exceptions no stratigraphy is actually available to date the walls in their own right, and therefore this dating is essentially based on the fact that all such places make their first appearance into *written* history through the works of the Roman historians (for instance, Livius) who mention the "deduction" of a Roman colony in the same sites. It is, therefore, assumed that the Romans were responsible for the construction of the walls after the conquest of the towns. Before the expansion of the Roman control, however, the ethnic scenario in central Italy was extremely complicated: the Romans were indeed only one among several Latin tribes, and the region was inhabited by many other people as well, such as the Hernics and the Volsceans, each one with their own culture, in active cultural and trade exchanges (or war) with the Latins, the Etruscans and the Mediterranean area. Thus, most, if not all, the settlements pre-date the Roman period, and the dating of their walls is actually uncertain, leaving us with the possibility that the original layout of the towns was conceived before the "orthogonal grid rule", during the first centuries of the Iron Age (from the ninth century onward).

In what follows, we shall investigate this possibility, considering, as said, four cities as case-studies. I hope, however, to extend this analysis in the future to many other towns in which the "topographical stratigraphy" is more complicated and deserves further investigation (towns which certainly deserve further attention are, for instance, Segni, Amelia and Alba Fucens). Our four "laboratories" here will be two Hernic towns which have been continuously inhabited up to the present, Ferentino and Alatri, and two towns whose original foundation is not certain (current opinion is that it is of early Roman age), which were already abandoned in Roman times, Norba and Cosa.

3.1 Ferentino

Ferentino was certainly inhabited since the seventh century B.C., perhaps earlier. Nobody, however, knows when the town walls were first constructed, although many archaeologists place them fully in the Roman period [Quilici and Quilici Gigli 1994]. The walls are in any case virtually intact, and show, superimposed over the megalithic structure, an elevation made out of squared blocks which is certainly Roman (fourth-third century B.C.) (fig. 6).

The same can be seen on the Acropolis, which is an imposing building constructed on a megalithic basement that is fifteen meters high (fig. 7).

The plan of the town can be seen in fig. 8a. It has *five* main gates. Among them, however, only *three* gates, A1-A2-A3, correspond to the original layout (in particular, the most important gate, denoted by A1, is the famous *Porta Sanguinaria*, shown in fig. 6), while gates B1 and B2 were added in late Republican times (end of the second century B.C.). It is therefore clear that the original town planning – whether it was Roman or not – was conceived on the basis of a *tripartite* structure. Connecting the three original gates one obtains a isosceles triangle (fig. 8b); it can be readily seen that the Acropolis lies in the center of this triangle.

The city plan was, therefore, based on just one main axis, oriented roughly north-south; when the Romans, after the conquest, decided to adapt the urban plan to their "squared" mentality, this axis became the Cardus of the city. They then opened the two new gates and the street – roughly oriented east-west – which connects them, playing the role of the Decumanus. The ideal, commercial, and social center of the city then became the point of intersection between the two main axes, and at this point was built the forum. What happened is thus very clear: a triangular city with the Acropolis at the ideal center was re-modelled as a "castrum" city with its center in the Forum. Interestingly enough, during the medieval age the Ferentino cathedral was built on the pre-existing Acropolis. As a consequence, the symmetry of the Roman *castrum* was broken again, since the cathedral was the "center" of the medieval social life. The people thus felt – perhaps unconsciously – the necessity of "restoring" the original symmetry of the town, and it was probably for this reason that two new churches were constructed near the two old city gates A2 and A3. The new churches, together with the main gate of the city, somewhat reconstructed the triangular symmetry [Montuori 1996].

Fig. 6. Ferentino: the south-east sector of the polygonal walls with the gate called *Porta Sanguinaria*

Fig. 7. Ferentino: the south-west bastion of the Acropolis

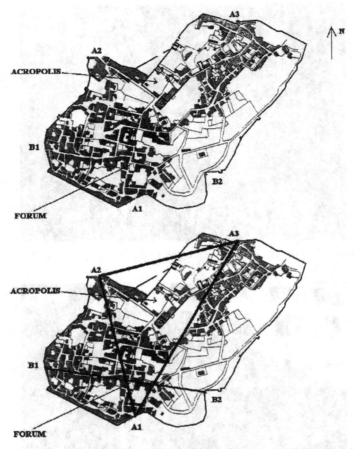

Fig. 8. a, above) Plan of Ferentino; b, below) Plan of Ferentino with the original
triangular layout indicated

3.2 Alatri

Among the cyclopean towns in Italy, Alatri is perhaps the most enigmatic. The city was
built around a small hill, and the town was surrounded by megalithic walls, of which many
remains are still visible today. The Acropolis is placed on the hill, and it is a gigantic
construction, a sort of huge "geometric castle" dominating the center of the town; although
generally not well known, it is one of the most impressive megalithic buildings in the entire
world, and the famous German historian Gregorovius (1821-1891) reported that it made
"an impression greater than that made by the Coliseum" (fig. 9). On the top of the hill
there existed a second megalithic structure, perhaps a palace or a temple, whose giant
basement was used in the Middle Age as the foundation for the Cathedral (some of the
blocks still visible on the right end side of the church reach dimensions of the order of 2 x 2
x 1.5 meters and weigh around 12 tons; in spite of this they are perfectly cut and joined
together at several – up to nine – corners) (fig. 10).

Fig. 9. Alatri: the north-west sector of the megalithic walls of the Acropolis

Fig. 10. Alatri: the point O, the megalithic basement and the *mundus*

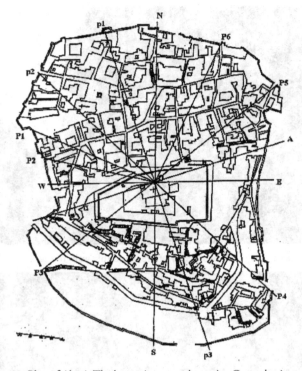

Fig. 11. Plan of Alatri. The layout is centered at point O, on the Acropolis

It is already well known that Alatri and its Acropolis were planned in accordance with rigorous mathematical and astronomical references [Capone 1982; Aveni and Capone 1985; Magli 2006b] and, in particular, it has been shown that the original layout of the city is based on a radial geometry. The center lies on the Acropolis, near the megalithic basement (in the point indicated by O in fig. 11), which therefore plays the role of ideal "focus" of the town.

The walls have six main gates (indicated by P1-P6) and three posterns (p1-p3). All the main gates, excluding P2, are equidistant from O, and therefore lie on a circle centered in O. The radius of the circle equals three times the value of the segment OH (which is about 92 m long) connecting the center with the north-east corner of the Acropolis and indicating the summer solstice sunrise. In addition, the town also shows a sort of curious quadripartite "symmetry". Indeed, the north-west sector has two main doors and two small ones, while the north-east sector two main doors. "Symmetrically dividing by two" with respect to O, the south-west sector has one main door and the south-east sector one main door and one small door. It is difficult to attribute all this to chance, because the lines connecting the pairs of doors, p1-p3, p2-P4, P3-P5, all intersect each other in O. Finally, near this point is visible a narrow and deep cleft in the rocks, which perhaps indicated the *mundus* of the city.

There can be little doubt about the fact that such a complex geometric layout predates the Roman period. Interestingly enough, archaeological excavations have shown that the Roman forum of the city was situated where the main square is located today, in the northern part of the town. Since in Alatri it was simply impossible to re-convert the urbanistic design to the "squared" conception, because of the presence of the huge hill of the Acropolis at the very center, the "social center" of the town was translated to the northern sector, while the southern one remained a residential quarter without centers of social aggregation, a role which is still preserved today [Ritarossi 1999].

3.3 Norba

Norba lies at the very end of the Lepini mountains, on a steep ridge which looks towards the sea, some 80 km south of Rome. The area surrounding the city was inhabited at least from the fourth century B.C., and, as in Ferentino, at Norba there are strong hints pointing to a peopling of the town itself at least from the eighth-seventh century B.C., although, again, the town walls are usually attributed to the Romans and dated to the fourth century B.C. [Quilici and Quilici Gigli 2001].

Fig. 12. Norba: the south-west front of the ramparts.

Fig. 13. Norba, the main gate or *Porta Maggiore*

The city is very big, with a perimeter of some 3 km, and the megalithic walls comprise three small hills, or Acropoli, each one with temples built on the top. The town was besieged and all its habitants killed during the Mario-Silla war (around 82 B.C.); since then the site was never again inhabited, so today it is one of the most beautifully preserved cyclopean towns of Central Italy (figs. 12,13).

The original layout of Norba was clearly inspired by the number 3: there are indeed three Acropoli (small hills) and, originally, three main gates A1, A2, A3 (fig. 14a). The contour of the walls was vaguely circular, but on the south-west side the builders strictly followed very steep cliffs, giving the town quite irregular boundaries. The interior of the city exhibits a rigid orthogonal grid which, however, at least in my opinion, cannot be contemporary with the construction of the walls but must be more recent. Indeed, the spectacular main gate on the east side, *Porta Maggiore* (fig. 13), does not lead to any of the east-west axes of the grid. Conversely, the paved Decumanus, which still today crosses the entire town, leads not to a gate but to the hill (called *Acropoli Minore*) located to the south of the main gate; at the other end the Decumanus enters into the city through gate B1, which in turn was almost certainly added after the initial construction of the walls [Quilici Gigli 2003]. The whole internal layout of the town is therefore attributable to a re-organization of the urban space made by the Romans in the second century B.C., and at this time the east-west axis became the main axis of the city (fig. 14b).

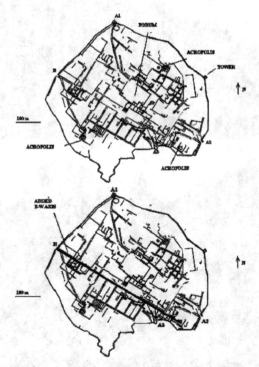

Fig. 14. a, above) Plan of Norba; b, below) Plan of Norba with the east-west axis highlighted

The view from the Decumanus was (and still is) really stunning: looking out, the eye goes to the horizon, while, looking in, one is directly in front of the spectacular ramp that ascends to the two temples on the top of the hill. One of them is parallel to the axis, while the orthogonal one was conceived to be viewed from the exterior of the town. Interestingly enough, the guideline of the Decumanus was fixed in accordance to an astronomical alignment: it is indeed easy to check that it points to the summer solstice sunset.

3.4 Cosa

The city of Cosa lies directly on the sea, on the Ansedonia promontory in southern Tuscany. The position of the town, high on the promontory, appears to be ideal for control of the sea behind, which was very important from the commercial point of view, due to the mines which are present on the coast above, from the Argentario peninsula to the island of Elba.

At the base of the hill, at less than 3 km from the town existed an ancient port, equipped with complex artificial structures. In particular, an impressive channel, the *Tagliata Etrusca*, still visible today, is carved out of the rocks; eighty meters long, twelve meters high and two meters wide, it was probably constructed as part of the works required for managing the port. However, this work is not mentioned in written sources, and nobody really knows for certain who built it, when, and why. Although a city named Cosa is mentioned as an Etruscan settlement by several ancient authors, including Virgil, today most archaeologists believe that the town was entirely a Roman colony, founded from scratch in the first half of the third century B.C. [Brown 1951, 1980]. The city was abandoned in the early Imperial age and therefore, like Norba, it appears as it was two thousand years ago.

The walls of Cosa are masterpieces of polygonal masonry (figs. 15,16) and are equipped (the only example in Italy) with several towers. These, however, were probably added in later times with respect to the original construction, since there is no joint between the blocks of the walls and those of the towers.

Fig. 15. Cosa: the north-west gate   Fig. 16. Cosa: the north-east gate

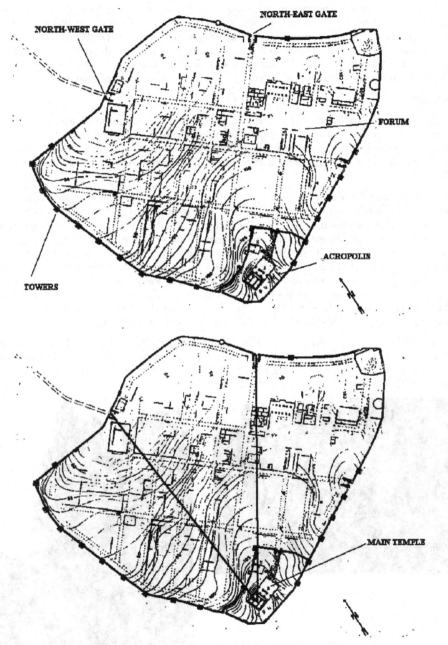

Fig. 17. a, above) Plan of Cosa; b, below) Plan of Cosa: two solid lines have been drawn to connect the *mundus* of the city, in the central cell of the Capitolium, with the two north gates

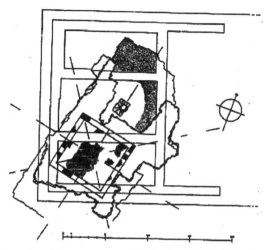

Fig. 18. Cosa: plan of the structures under the Capitolium (after [Brown 1980])

The interior of the city was planned on the basis of a rigid orthogonal grid, and this grid is in accordance with the disposition of the gates, so that – although no internal axis can be defined Cardus or Decumanus, because none of them connects two gates – the hippodamean layout might reasonably be considered contemporary with the construction of the walls (fig. 17a). Curiously enough, however, it seems that the design was inspired from its very conception by a *tripartite* division of the urban space. First of all, the city had only three main gates (further to these, there is only one postern, located near the Acropolis). Second, during the excavation of the main temple on the Acropolis, the Capitolium dedicated to the *Triade Capitolina*, a squared basement, roughly oriented to the cardinal points, was discovered (fig. 18). The basement is of course more ancient than the temple above it, and it almost certainly refers to the first phase of construction of the town. At a few meters behind the basement, on axis with it and at the very center of the temple's cells, a natural rocky cleft was found; this probably contained a foundation deposit of first produce [Brown 1980]. Thus, the archaeologists probably found in this the *mundus* of the city, but *it does not correspond to a geometrical center of the town's orthogonal grid*, since the Acropolis lies in a spectacular position at the southern corner of the city walls, dominating the sea behind. Thus, how should it be interpreted? If we ignore for a moment the orthogonal grid on which the streets of Cosa were laid out, the *mundus* actually turns out to have a geometrical function: the lines connecting it with the two northern doors divide the city in *three* quarters which are very similar in size, and, in addition, the line pointing to the north-west gate is oriented on the meridian (fig. 17b).

4 Discussion and conclusions

We have thus seen that the layouts of four of the ancient towns of central Italy show urbanistic features based on the number 3 (three gates, three Acropoli, tripartite or even radial division of urban space) which can hardly be attributed to the Roman period (or, at least, they do not correspond to what we know about this period). These features appear to make reference to an older tradition, one perhaps contained in the lost Etruscan books. As a matter of fact, the Piacenza Liver is radially divided and exhibits three "hills", and Servius states explicitly that the rules for the foundation of cities, which were contained in the books, were governed the number 3. If this tradition really existed, it would have pre-dated the period in which the orthogonal grid became the rule, around the early sixth century. This proposal is independently supported by other data, such as astronomical alignments (see also [Magli 2006b, 2006c]); and, in view of the new excavations on the Palatino hill, we know that the first fortification of Rome itself must also be retro-dated from the standard period – beginning of the sixth century – to the *traditional* one indicated by the Roman historians, around the middle of the eighth century B.C. (by the way, due to the enormous amount of archaeological stratifications, the layout of the archaic town of Rome remains uncertain, see [Carandini 1997]).

A problem now arises, namely, how are we to understand in which cultural horizon the "radial" or the "tripartite" geometrical planning should be collocated? It is possible that this tradition originated in Greece. In fact, as mentioned in the introduction to the present paper, although *no* classical Greek city was ever laid out radially, there are some enigmatic passages by Greek writers that mention a radially planned town [Cahill 2002]. First of all, one could cite Plato's famous description of Atlantis in the *Critia*, in which a circular town surrounded by concentric water channels is depicted. Secondly, leaving aside the many controversial questions about this "ideal" city, in his last dialogue, *The Laws* (written around 460 B.C.), Plato states the way in which *all* new cities should be planned by saying:

> We will divide the city into twelve portions, first founding temples to Hestia, to Zeus and to Athena, in a spot which we will call the Acropolis, and surround it with a circular wall, making the division of the entire city and country radiate from this point (transl. Benjamin Jowett).

A star-like town also appears in the comedy *Birds*, written by Aristophanes around 415 B.C. In this comedy, a person called Meton (probably a caricature of the astronomer Meton of Athens) proposes planning a city in this way:

> With the straight ruler I set to work to inscribe a square within this circle; in its center will be the market-place, into which all the straight streets will lead, converging to this center like a star, which, although only orbicular, sends forth its rays in a straight line from all sides (anonymous transl., 1917).

The passage is clearly satiric but, even if Aristophanes was attempting to *criticize* Meton (an intention which is far from clear), in any case he depicts an urban plan which is, once again, star-like. This "theoretical" star-like town has generated much debate, and there is no satisfactory interpretation available. For instance, the authoritative scholar Francesco Castagnoli wrote:

> If the comparison is taken literally, the vision of a "Place de l'Étoile" arises. But such a plan was not employed until the seventeenth century; it was totally unknown to the ancient world. Though it is true that the poet can create before the architect, a less literal interpretation of the passage would be appropriate: the rays are the four streets which, spreading from the agora, define the quadripartite city.

Thus, the "rays" should be the four main streets of the squared town, an interpretation which is at least questionable. However, it is not true that a radial plan was totally unknown to the *ancient* world: it was unknown (or, better, unrealised) in the *classical* world, but the radial organization of the inhabited space *is* present in Greece before the classical period: it is indeed attested to, for instance, by the Neolithic site of Dimini (fig. 19) and in the Mycenaean citadel of Aghios Georgios near Chalandritsa; further, it should be noticed that the radial planning of settlements was the rule in Palestine during the Iron Age [Finkelstein 1988]; see, for example, the layout of the site of Nasbeh, dating around the ninth century B.C. (fig. 20).

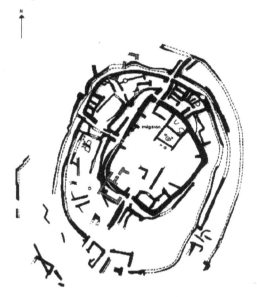

Fig. 19. Plan of the Neolithic settlement of Dimini

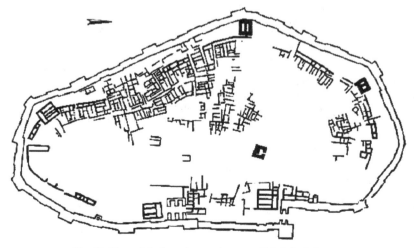

Fig. 20. Plan of the Iron Age settlement of Nasbeh, Palestine

Therefore, it might be that Plato and Aristophanes, like Servius, were referring to old Mediterranean traditions, filtered during the Hellenic Middle Age, and thus it might well be that the building techniques and the layouts which are visible in the Italian cyclopean towns originated in Greece, during the Hellenic Dark Age, or, even before, in the Mycenaean period.

Note added in proofs:

After the completion of this work, the author became aware of a new, very important discovery in southern Lazio.

Indeed, archaeologists Lorenzo Quilici and Stefania Quilici Gigli have discovered the remains of a previously-unnoticed archaic town located on the steep hill called Pianara, near Fondi. This town, certainly inhabited from the sixth to the fourth centuries B.C., is probably the one called "Amyclae" by the Roman historians. The town is fortified with a imposing polygonal wall, and it turns out that the circuit of the walls comprises three hills and three main gates.

Notes

1. *Miratur molem Aeneas, magalia quondam – miratur portas strepitumque et strata viarum*, which roughly means "Look at the size (of the town) Enea, where before was only rubbish – look at the gates, the street traffic and the industrious people."

2. *Prudentes Etruscae disciplinae aiunt apud conditores Etruscarum urbium non putatas iustas urbes, in quibus non tres portae essent dedicatae et tot viae et tot templa, Iovis Iunonis Minervae.*

References

AVENI A. and G. CAPONE. 1985. Possible Astronomical Reference in the Urbanistic Design of Ancient Alatri, Lazio, Italy. *Archaeoastronomy* 8: 12.

AVENI, A. and G. ROMANO. 1994. Orientation and Etruscan Ritual. *Antiquity* **68**: 545-563.

BARALE, P., M. CODEBO, and H. DE SANTIS. 2002. *Augusta Bagiennorum (Regio Ix) Una Città Astronomicamente Orientata*. Turin: Ed. Centro Studi Piemontesi.

BLOCH, R. 1970. Urbanisme et religion chez les etrusques: explication d'un passage fameux de Servius. In *Studi sulla citta' antica : atti del convegno di studi sulla citta' etrusca e italica preromana*. Bologna: Istituto per la storia di Bologna.

BONGHI-JOVINO, M. 1998. Tarquinia: Riflessioni sugli interventi tra metodologia, prassi e problemi di interpretazione storica. Pp. 41-51 in *Archeologia della città. Quindici anni di scavo a Tarquinia*, M. Bonghi Jovino, ed. Milan.

———. 2000. Il complesso sacro-istituzionale di Tarquinia. Pp. 265-270 in *Roma, Romolo, Remo e la fondazione della città*, A. Carandini and R. Cappelli, eds. Rome: Electa.

BROWN, F. 1951. Cosa I. History and Topography. *Memories of the American Academy in Rome* **20.**

———. 1980. *Cosa: the making of a Roman Town*. Ann Arbor: University of Michigan Press.

CAHILL. N. 2002. *Household and City Organization at Olynthus*. New Haven: Yale University Press.

CAPONE, G. 1982. *La progenie hetea*. Alatri: Tofani ed.

CARANDINI, A. 1997. *La nascita di Roma*. Turin: Einaudi.

CASTAGNOLI, F. 1971. *Orthogonal town planning in antiquity*. Cambridge, MA: MIT press.

FINKELSTEIN, I. 1988. *The Archaeology of the Israelite Settlement*. Jerusalem: Brill Academic.

INCERTI, M. 1999. The Urban fabric of Bologna: orientation problems. Pp. 3-12 in *Atti*, VI international seminar on urban form, R. Corona and G.L. Maffei, eds., Florence: Universita' degli Studi di Firenze.

LIBERATORE, D. 2004. *Alba Fucens: Studi di storia e di topografia*. Rome: L'Erma di Bretschneider.

LUGLI, G. 1957. *La tecnica edilizia romana con particolare riguardo a Roma e al Lazio.* Rome: Bardi.

MAGLI, G. 2006a. Mathematics, Astronomy and Sacred Landscape in the Inka Heartland. *Nexus Network Journal* 7, 2: 22-32.

————. 2006b. The Acropolis of Alatri: astronomy and architecture. *Nexus Network Journal* 8, 1: 5-16.

————. 2006c. Polygonal walls in the Latium Vetus: an archaeo-astronomical approach. In *Proceedings*, SEAC 2006 conference, J. Liritzis et al., eds. *Mediterranean Archaeology and Archaeometry* 6, in press.

MANSUELLI, G.A. 1965. Contributo allo studio dell'urbanistica di Marzabotto. *Parola del passato : Rivista di studi antichi* XX: 314.

MERTENS, J. 1969. *Alba Fucens. I-II.* Rapports et études. Brussells and Rome.

MONTUORI, F. 1996. Ferentino: Piano di recupero della cinta muraria. In *Cinte murarie di antiche città del Lazio.* Comm. Eur.- Progetto Raphael. S. Quirico d'Orcia: Editrice Don Chisciotte.

PALLOTTINO, M. 1997. *Etruscologia.* Milan: Hoepli.

QUILICI, L. and QUILICI GIGLI, S. 1994. Ricerca topografica a Ferentinum. Pp. 159-254 in *Opere di assetto territoriale ed urbano.* Atlante tematico di topografia antica, vol. 3. Rome: L'Erma di Bretschneider.

————. 2001. Sulle mura di Norba. Pp. 181-244 in *Fortificazioni antiche in Italia: età repubblicana.* Atlante tematico di topografia antica, vol. 9. Rome : L'Erma di Bretschneider.

QUILICI GIGLI, S. 2003. Norba: l'acropoli minore e i suoi templi. Pp. 289-321 in *Santuari e luoghi di culto nell'Italia antica.* Atlante tematico di topografia antica, vol. 12. Rome: L'Erma di Bretschneider.

RITAROSSI, M. 1999. *Aletrium.* Alatri: Tofani editore.

RYKWERT, J. 1999. *The Idea of a Town: The Anthropology of Urban Form in Rome, Italy, and The Ancient World.* Cambridge, MA: MIT Press.

SOUTHWORTH M. and E. Ben-Joseph. 1997. *Streets and the Shaping of Towns and Cities.* New York: McGraw-Hill.

TORELLI, M. 1966. Un templum augurale d'età repubblicana a Bastia. *Rendiconti dell'Accademia nazionale dei Lincei – Classe Sc. morali storiche filologiche* XXI: 293-315.

About the author

Giulio Magli is a full professor of Mathematical Physics at the Faculty of Civil, Environmental and Land Planning Engineering of the Politecnico of Milan, where he teaches courses on Differential Equations and Rational Mechanics. He earned a Ph.D. in Mathematics at the University of Milan in 1992 and his research activity developed mainly in the field of General Relativity Theory, with special attention to problems of relevance in Astrophysics, such as stellar collapse. His research interests however include History of Astronomy and Archaeoastronomy, with special emphasis on the relationship between architecture, landscape and the astronomical lore of ancient cultures. On this subject he recently authored the book *Mysteries and Discoveries of Archaeoastronomy*, published in 2005 (in Italian) by Newton & Compton.

James Harris

25 West Drive
Plandome, N.Y. 11030 USA
jharris@related.com

Keywords: algorithms, fractals,
rule-based architecture, Frank
Lloyd Wright, Piet Mondrian

Research

Integrated Function Systems and Organic Architecture from Wright to Mondrian

Abstract.. The development of an architectural form where the individual parts reflect the integrated whole has been a design goal from ancient architecture to the current explorations into self-organizational structures. Organic architecture, with this part-to-whole association as an element of its foundation, has been explored from its incidental use in vernacular structures to its conscious endorsement by Frank Lloyd Wright. Traditionally Piet Mondrian has not been associated with organic architecture but a closer examination of the artistic and philosophical underpinnings of his work reveals a conceptual connection with organic architecture.

The development of an architectural form where the individual parts reflect the integrated whole has been a design goal from ancient architecture to the current explorations into self-organizational structures. From medieval castles such as the Castel de Monte, Gothic cathedrals such Reims to Hindu temples [Sala 2000] historically it appears that our minds are oriented to appreciate buildings constructed with this quality [Salingaros 2001]. Organic architecture, with this part-to-whole association as an element of its foundation, has been explored from its incidental use in vernacular structures to its conscious endorsement by Frank Lloyd Wright. Traditionally Piet Mondrian has not been associated with organic architecture but a closer examination of the artistic and philosophical underpinnings of his work reveals a conceptual connection with organic architecture. This relationship is explored through the application of some of nature's fundamental structural principles to his artistic style.

The appreciation of beauty is one of the most basic human capacities. Architectural beauty is developed by the awareness of a balance of order in diversity of the unified whole and its constituent parts or the "ensemble effect of beauty"[Langhein 2001]. The quest to understand beauty in architectural design leads us to examine the intrinsic idea of nature [Bovill 1996].

Gestalt psychology focuses on visual perception to emphasize the dynamic interplay of parts and whole. Gestalt is usually translated as form or organized structure and is rooted in German thought of the self-actualizing wholeness of organic forms [Hubert]. It holds that we have certain tendencies to perceive visual data in organized or configurational terms [Detrie 2002] and to "constellate" or to see as "belonging together" elements that look alike, are proximate to each other, are similarly spaced, or are arranged in such a way that they appear to continue each other. The appearance of parts is determined by and understood relative to the systematic whole [Behrens; Crowe 1999]. A Gestalt quality is not concerned with the combination of the various elements per se but in the entity that is created based on their unity but is still discernible from them [Lyons].

Gestalt theorists believe in an aesthetic dimension of inherent order in nature [Lyons]. Nature has a highly complex and ordered system which we connect to through our involuntary and subconscious perceptual system as well as our conscious understanding. As we are a part of nature's province it is logical that we are structured to appreciate nature's underpinnings [Detrie 2002]. Our constructed world is an alternative environment and it is inherent that there is a coherent connection, subconscious or conscious, between it and the natural world. In the striving for a harmony between the two worlds it is compelling to search for the source of nature's structure.

Jean Piaget, the noted cognitive developmental psychologist, advanced a theory of structure based on three properties: wholeness, transformation and self regulation. Wholeness, the "defining mark of structures", developed from elements that were subordinated to laws and it is in terms of these laws that the structural whole is defined. Transformations are laws that govern the structure's composition and are based on the overall properties of the whole. Self regulation is the concept that the transformations tend to develop elements that belong to the structure and preserve its laws [Kranbuehl 2000]. All of these properties bear the hallmarks of Integrated Function System (IFS) fractals.

In the 1950s Beniot Mandlebrot formalized the study of fractals, which had been ongoing since the nineteenth century, culminating in his book *The Fractal Geometry of Nature* [1977]. Due to the amount of calculations involved, this study was stunted until the advent of computers. Mandelbrot's goal was to describe nature with geometry and numbers in order to illuminate its underlying structure. There are a number of different types of fractals, including escape time fractals, which are the source of many fractal images. Fractals based on IFS have been shown to possess nature's structural traits. Fractals have been cited as being representative of real plant structures, among other natural structures, whose distinctive feature is self-similarity based on a recursive procedure called algorithms for creating these forms. The self-similarity trait manifests itself in the analysis of the fractal on different scaling levels. An IFS fractal is produced by taking a starting object, such as a box, as a seed shape or initiator. Copies of the seed shape are manipulated by certain permitted affine transformations, such as rotation, translation, shearing, and scaling. The summation of these transformed copies becomes the new seed shape, which is again transformed by the exact same set of transformations that was used to create it. Each set of transformations is called an iteration. As the amount of iterations increase the initial seed shape becomes less evident and the rules used to create the subsequent seed shapes become more significant. The transformation rules, which are made up of the various affine transformations, are the essence of the fractal form. The fractal rules are expressed in the following format, which can be viewed with a text editor (Table 1).

a	b	c	d	e	f	g	h	i	j	k	l	p
0	0	0	0	0	0	0	0	0	0	0	0	0
0	0	0	0	0	0	0	0	0	0	0	0	0
0	0	0	0	0	0	0	0	0	0	0	0	0
0	0	0	0	0	0	0	0	0	0	0	0	0

Table 1.

In this example there would be a set of four transformed copies. The letters a, b, c, d, e, f, g, h and i are values which describe the rotation, shearing and scaling transformations on the object in three dimensional space; j, k, and l are the vectors which describe three-dimensional translation. P is a probability factor and the summation of that column will equal 1. The critical determinant elements in an IFS fractal structure are the affine transformations or rules represented by these values.

A classic IFS fractal is a tree. In fig. 1, the seed shape is shown on the left as a slightly deformed column. To the right of it is the graphic representation of the transformation rules which constitute the first iteration. The rules are three-fold: 1) take the seed shape, reduce it, rotate it slightly to the right, and translate it approximately halfway up the seed shape; 2) take the seed shape, reduce it, rotate it slightly to the left, and translate it approximately three quarters up the seed shape; 3) take the seed shape, reduce it, rotate it slightly to the right, and translate it close to the top of the seed shape. The figure at the right represents the fractal after three iterations.

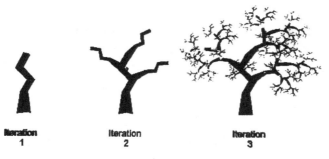

Fig. 1

The component parts of an IFS fractal after the first iteration look like a scaled-down copy of the whole and demonstrates the self-similar character of fractals in general. As demonstrated in fig. 1 and the table above, IFS fractals contain the "DNA" structure of natural objects and form the connection between mathematical elements and nature. Compositions based on IFS fractals can be the source of organic architecture whose inspiration is the underlying structure of a fractal [Lorenz 2003].

Frank Lloyd Wright believed in the importance of the study of nature as the basis for establishing an emerging American architecture. In a speech to the Royal Institute of British Architects he declared, "Modern architecture is a natural architecture—the architecture of nature, for Nature" [Wright 1939, 10]. Wright's mentor, Louis Sullivan, studied botanist Asa Gray and utilized the "manipulation of the organic" in the development of motifs [Gans and Kuz 2003]. Wright used nature as the basis of his geometrical abstraction and wanted to extract the geometry he found in nature [Eaton 1998]. He held that "nature means the essential significant life of the thing" and believed the term organic meant the relationship "in which the part is to the whole as the whole is to the part and which all devoted to a purpose consistently...some connection with this inner thing called the law of nature" [Meehan 1987]. These relationships are the essence of IFS forms and the basis of natural structures.

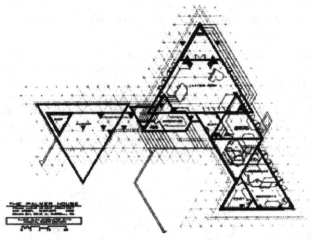

Fig. 2

One project of Wright's that has been cited for its fractal qualities is the Palmer House [Eaton 1998] (fig. 2). The citation of this building points out the misconception that a repetition of a form, the triangle in this case, constitutes a fractal quality. It is not the repetition of the form or motif but the manner in which it is repeated or its structure and nesting characteristics which are important. In fig. 3 I have developed a fractal form which is similar to the Palmer House plan. In the first iteration, the seed shape is shown in darker gray and the three transformations are shown in lighter gray. In the second and third iterations the original seed shape is dropped out and the iterations are color coded in three shades of gray.

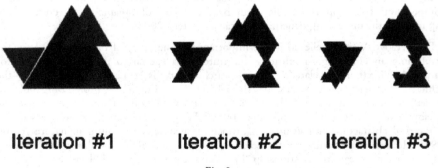

Iteration #1   Iteration #2   Iteration #3

Fig. 3

Fig. 4. Photograph by Thomas A. Heinz, AIA
© Thomas A. Heinz

Fig. 5

Frank Lloyd Wright only had two high-rises built: the Johnson Wax building in Racine Wisconsin and the Price Tower in Bartlesville, Oklahoma (fig. 4). In fig. 5 I have developed an IFS fractal model, utilizing 3D Max Maxscript, which approximates the Price Tower. The seed shape is an extruded square which represents the shaft of the Price Tower. The transformation rules consist of scaling, rotations and skewing of the seed shape and can be categorized into the base, shaft and top portions of the building. The bottom portion consists of four seed shapes scaled down to small vertical slabs with four additional copies of the seed shape scaled down to bars. These scaled-down seed shapes are then rotated for a pinwheel effect in plan. An additional scaled-down seed shape representing the lobby is skewed and translated to the corner. The middle portion consists of three sets of groupings stacked vertically. Each grouping consists of a spandrel level, a vision glass level and two sets of horizontal mullions. Each of these was a vertically scaled-down version of the seed shape. At the spandrel and vision glass levels four reduced copies of the seed shape are skewed and rotated as per the pinwheel structure shown in plan and another reduced copy of the seed shape straddles one corner of the seed shape. The crown portion, similar to the base area, has four seed shapes scaled down to thin slabs, rotated to produce a pinwheel shape similar to the base. There is a scaled-down seed shape that remains in the center and another reduced copy of the seed shape translated over to where it straddles one corner of the seed shape similar to the shaft portion of the structure. Lastly there are two copies of the seed shape which are skewed and rotated as one arm of the pinwheel. They are arranged above each other to continue the vertical arrangement from the middle section.

Wright's buildings have been reviewed and appreciated by critics for exhibiting "fractal" qualities at different scales. As one approaches one of his structures there is a progression or unfolding of additional elements or details which reflect variations of buildings characteristics [Lorenz 2003]. Although this experience is not a direct correspondence to the self-similarity characteristic of IFS based fractals, it is analogous; this is especially interesting considering that the concept of fractals did not become wide-spread until years after Wright's death.

Norman Crowe compared Wright and Le Corbusier, contemporaries of Mondrian, and the relationship of their architecture to nature [Crowe 1999]. A comparison of their respective masterpieces, Fallingwater and Villa Savoye, highlights the differences in their approach. While Villa Savoye represents a "celestial vision" developed out of Greco-Roman classicism, Fallingwater brings the visitor back to earth with a structure based on an emerging American architecture enmeshed in its natural setting. Cynthia Schneider has compared Fallingwater with another example of celestial architecture, the Guggenheim Museum in New York [Schneider 1999]. The Guggenheim interacts with its environment in a way similar to the Villa Savoye but, like Fallingwater, it is considered an example of organic architecture. Thorsten Schnier and John Gero have used principles of genetic engineering on examples of Wright's stained glass and Mondrian's artwork [Schnier and Gero 1998]. They performed the genetic operations of mutation and cross-over of their respective genes to create hybrid art forms. These genes are analogous to the IFS codes that map the transformations of organic fractal architecture.

Wright and Mondrian believed that Truth could be achieved by reducing nature to her most fundamental forms. They viewed their art as part of the dynamic whole of the cosmos expressing inner harmonies and the basic truths of the universe. These views are characteristic of the reduction of nature's forms to IFS codes and the fractal relation of the part to the whole.

Mondrian was a disciple of Theosophy and believed religion and art were on parallel paths: the aim of both was to transcend matter and understand the universal. "Religion always sought to harmonize man with nature, that is, with untransformed nature" [Mondrian 1993]. Mondrian revealed his relationship with nature in his essay "Natural Reality and Abstract Reality," in which his characters agree that beauty and inspiration are to be found in nature. The Mondrian character, the Abstract-Real Painter, believes they need in a sense to look through nature for its transcendental knowledge. He wanted to pierce the chaotic complexity of natural appearances, to glimpse past this veil to see the ultimate harmony and "cosmic rhythm" of the universe. The aim of his art was to capture the underlying structure of nature. Given the enduring popularity of his work "perhaps Mondrian succeeded in glimpsing through nature's veil with unmatched clarity" [Taylor 2004]. The relationship of Mondrian's art to the underlying transcendental structure of nature makes it an appealing candidate for applications of IFS.

Although Mondrian's art is associated only with two-dimensional representation, he frequently wrote about architecture and Neo-Plasticism. In his article "Toward the True Vision of Reality" Mondrian noted "While Neo-Plasticism now has its own intrinsic value, as painting and sculpture, it may be considered as a preparation for future architecture" [Mondrian 1993]. In "Home-Street-City" he wrote, "The application of Neo-Plastic laws is the path of progress in architecture" [Mondrian 1993] and in "Is Painting Inferior to Architecture?" he stated, "The new aesthetic for architecture is that of new painting....Through the unity of the new aesthetic, architecture and painting can together form one art and can dissolve each other" [Mondrian 1993]. He experimented with architectural aspects of his work in some stage designs as well as the design of his own studio. He was part of the De Stijl movement, which significantly affected architectural design with some notable buildings.

Mondrian based his art on the presence of scaffolds and the spatial interplay between their pictorial parts, which is analogous to the manifestation of IFS rules or codes and the resulting interplay of the parts to the whole. It seems fitting to explore this relationship by overlaying IFS with Mondrian's work. In fig. 6, we take the slab seed shape and apply transformation rules to produce a first iteration of a building facade that is inspired by Mondrian's *Composition No.1, Composition with Yellow*. The upper right element and the bottom three elements, in addition to being scaled down, are rotated 90° to the right. The middle element on the left side is scaled down and rotated 180°. The upper left element is simply scaled down. In the second and third iterations (figs. 7 and 8), they increasingly exhibit the rhythm Mondrian discussed in his writings and approach the vitality of his Boogie-Woogie paintings. In figs. 9 and 10 this facade is incorporated with another Mondrian-influenced IFS facade into an architectural structure.

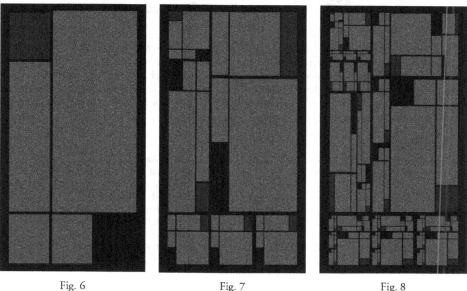

Fig. 6   Fig. 7   Fig. 8

Figs. 9, 10

I have been struck with the natural association of IFS and architecture to generate structures that engender a natural affinity in the observer. In applying IFS to Frank Lloyd Wright, the principal architect associated with organic architecture, and the artist Piet Mondrian, who represents the other end of the spectrum, I hope I have demonstrated the range of possible architectural applications of IFS. I have continued exploring this concept at http://www.fractalarchitect.com and encourage the reader to continue the experiment.

References

BOVILL, Carl. 1996. Fractal Geometry in Architecture and Design. Basel: Birkhäuser.

CROWE, Norman. 1999. *Nature and the Idea of a Man-Made World.* Cambridge MA: MIT Press.

DETRIE, Thomas. 2002. Gestalt Principles and Dynamic Symmetry: Nature's Connection to Our Built Environment. http://www.public.asu.edu/~detrie/msj.uc_daap/article.html

EATON, Leonard K. 1998. Fractal Geometry in the Late Work of Frank Lloyd Wright. Pp. 23-38 in *Nexus II: Architecture and Mathematics*, Kim Williams, ed. Fucecchio (Florence): Edizioni dell'Erba.

GANS, Deborah and Zebra KUZ, eds. 2003. *The Organic Approach to Architecture.* London: Wiley-Academy.

GOLDING, John. 2000. *Paths to the Absolute.* Princeton: Princeton University Press.

HUBERT, Christian. No date. Gestalt. http://www.christianhubert.com/hypertext/Gestalt.html

KRANBUEHL, Don. 2000. Interplay. Master's thesis, Virginia Polytechnic Institute and State University.
http://scholar.lib.vt.edu/theses/available/etd-03152000-14390004/unrestricted/interplay1-13B.pdf

LANGHEIN, Joachim. 2001. "INTBAU:Proportion and Traditional Architecture. *International Network for Traditional Building, Architecture and Urbanism (INTBAU)* 1, 10. http://www.intbau.org/essay10.htm.

LYONS, Andrew. No date. Gestalt Approaches to the Virtual Gesamtkunstwerk. http://www.users.bigpond.com/tstex/gestalt.htm.

LORENZ, Wolfgang. 2003. Fractals and Fractal Architecture. Master's thesis, Vienna University of Technology. http://www.iemar.tuwien.ac.at/modul23/Fractals/

MANDELBROT, Benoit. 1977. *The Fractal Geometry of Nature*. New York: W.H. Freeman.

MEEHAN, Patrick. 1987. *Truth Against the World*. Wiley-Interscience.

MONDRIAN, Piet. 1993. The Realization of Neo-Plasticism in the Distant Future and in Architecture Today. In *The New Art-The New Life: The Collected Writings of Piet Mondrian*. Harry Holtzman and Martin S. James, eds. Cambridge MA: Da Capo Press.

SALA, Nicoletta. 2000. Fractal Models in Architecture: A Case Study. In *Proceedings of the International Conference on "Mathematics for Living"*, Jordan, 18-23 November 2000. http://math.unipa.it/~grim/Jsalaworkshop.pdf.

SALINGAROS, Nikos A. 2001. Fractals in the New Architecture. *Archimagazine*. http://www.math.utsa.edu/sphere/salingar/fractals.html

SCHNEIDER, Cynthia. 1999. Frank Lloyd Wright Lecture. Amsterdam, 18 June 1999. http://hpi.georgetown.edu/lifesciandsociety/pdfs/franklloydwright061899.pdf.

SCHNIER, Thorsten and John GERO. 1998. From Mondrian to Frank Lloyd Wright: Transforming Evolving Representations. http://citeseer.ist.psu.edu/cache/papers/cs/1816/http:zSzzSzwww.arch.usyd.edu.auzSz~thorstenzSzpublicationszSzacdm98.pdf/from-frank-lloyd-wright.pdf.

TAYLOR, Richard. 2004. Pollock, Mondrian and Nature: Recent Scientific Investigations. *Chaos and Complexity Letters* **1**, 29.

WRIGHT, Frank Lloyd. 1939. *An Organic Architecture: The Architecture of Democracy*. Cambridge MA: MIT Press.

About the author

James Harris has a degree in architecture from Catholic University, an MBA from Fordham University and is currently a licensed architect in New York City. As a Senior Vice President for the Related Companies, he has been involved in large scale New York developments, building over four million square feet of commercial and residential structures in Manhattan.

Charoula
Stathopoulou

University of Thessaly
Papagou 30,
15343 Athens, Greece
stath@rhodes.aegean.gr

Keywords: Cognition, context,
culture, designing activity,
Ethnomathematics, informal
mathematics, teaching of
mathematics

Research

Traditional patterns in Pyrgi of Chios: Mathematics and Community

Abstract. Ethnomathematical research has revealed interesting artifacts in several cultures all around the world. Although the majority of them come from Africa, some interesting ones exist in Western cultures too. *Xysta* of Pyrgi are a designing tradition that concerns the construction of mainly geometrical patterns on building façades by scratching plaster. The history and the culture of the community, the way that this tradition is connected with them, as well as the informal mathematical ideas that are incorporated in this tradition are some of the issues that are explored here.

Η εθνομαθηματική έρευνα έχει φέρει στην επιφάνεια ενδιαφέροντα τεχνουργήματα από όλο τον κόσμο. Παρότι η πλειονότητά τους προέρχεται από την Αφρική ενδιαφέροντα σχέδια υπάρχουν και σε δυτικές κουλτούρες. Τα ξυστά στο Πυργί της Χίου αποτελούν μια σχεδιαστική παράδοση που αφορά στην κατασκευή γεωμετρικών, κυρίως, σχεδίων στο σοβά που βρίσκεται στην πρόσοψη των σπιτιών. Σ' αυτή την εργασία εξερευνάται κυρίως η ιστορία και η κουλτούρα της κοινότητας του Πυργιού, η σύνδεσή τους με την παράδοση των ξυστών καθώς και οι άτυπες μαθηματικές ιδέες που είναι ενσωματωμένες σ' αυτή την παράδοση.

1 Theoretical points

1.1. Introduction

According to D'Ambrosio, "Mathematics is an intellectual instrument created by the human species to help in resolving situations presented in everyday life and to describe and explain the real world" [2005, 11]. So, every community depending on its special environmental and social conditions—not necessarily practical—selects different ways to answer its own needs.

As Paulus Gerdes notes:

> Many peoples do not appear to have referred to the mathematics history books. This does not mean that these people have not produced mathematical ideas. It means only that their ideas have not (as yet) been recognised, understood or analysed by professional mathematicians and historians of mathematical knowledge. In this respect the role of Ethnomathematics as a research area resides in contributing with studies that permit to begin with the recognition of mathematical ideas of these people and to value their knowledge in diverse ways, including the use of this knowledge as a starting base in mathematics education [2005].

Ethnomathematical research has shown that all cultures use notions and practices recognizable as mathematical no matter whether or not mathematics exists as a distinguishable category of cognition in these cultures. Different cultures present and develop some common mathematical activities in order to respond to the needs and requirements of the natural as well as the sociocultural environment. That is to say, depending on the needs and demands, mathematical activities tend to be developed in different directions as well as in different degrees in every culture.

According to Bishop [1988] the mathematical activities that are accepted as universal are counting, measuring, locating, designing, playing, and explaining. These six activities are adopted as analytical categories by all researchers and very often research focuses on one of them.

Research regarding the above activities contributes to the acquisition of a deep cognition about mathematical activities and the ways through which people are educated by them in every particular culture. Also, it helps us to realise that all cultures have common characteristics as well as particular ones that distinguish them. Furthermore, the study of these universal activities results in the recognition and acceptance of each culture's contribution to what we today call academic or school mathematics.

1.2. Design activity

The activity of designing concerns "the manufactured objects, artifacts and technology which all cultures create for their home life, for trade, for adornment, for warfare, for games and for religious purposes" [Bishop 1988, 39]. An important part of designing concerns the transformation of some materials, usually from nature, into something that is useful in a given society with particular conditions.

Design activity exists in every culture. The type of designs depends on the people's needs and the available materials. What differs among cultures is what is designed, in what way and for what purpose. That is to say, in every society depending on its own needs—not always material—the expression of this particular activity is differentiated.

Some researchers, as for example Pinxten [1983], write about their own impression of the geometrical and mathematical possibilities of the design forms that appear in several cultures they have studied. In her book *Africa Counts* [1973], Zaslavsky presents the richly geometrical tradition of African societies, part of which is decorative patterns. She also describes the African architecture that is depicted on houses in the form of elaborate drawings.

Gay and Cole [1967] note that the Kpelle have developed a technology for the construction of houses using right angles and circles: "they know that if the opposite sides of a quadrilateral are of equal length and if the diagonals are also of equal length, the resulting figure will be a rectangle" [Bishop: 1988, 41]. The Kpelle, although unable to state this suggestion as a theorem, apply it in their constructions as a culturally acquired cognition.

Geometrical figures such as the right angle and the orthogonal triangle appear frequently in all cultures around the world. Circles also play an important role among symbolic representations, such as in mandalas. Several geometrical figures played important role in helping people to imagine relations between phenomena.

Paulus Gerdes [1996, 1999] gives various examples concerning mathematical ideas incorporated in the design processes of artists in Mozambique as well as in other places in Africa. Furthermore he emphasizes the necessity of incorporating this cognition into the curriculum. He maintains that if the hidden mathematics of Mozambique, which he characterizes as "frozen" mathematics, were "defrosted", the culture would be revealed and make it clear that Mozambique's people, like other people, have produced mathematics.

Among other interesting design traditions connected with mathematical ideas is that of the *quipu* of the Incas, studied by Marcia and Robert Asher [1981]. A quipu is an assemblage of coloured and knotted cotton cords. The colours of the cords, the way they are connected together, their relative placement, the spaces between them, the types of knots on the individual cords, and the relative placement of the knots are all part of a logical-numerical recording. In the tradition of the quipu exist important mathematical ideas, mostly of graph theory, which the Incas developed much earlier than did the West. Also, it is important to mention the fact that the representation of the numbers developed in a way that took into consideration the place value and the representation of zero.

Another interesting expression of design activity is found in the tradition of *sona*, the name given by the Tchokwe people of northeast Angola to their standardized drawings in sand. These were used as mnemonic aids in the narration of proverbs, fables, riddles, etc. Thus the patterns of *sona* played an important role in the community's transmission of collective memories. The *sona* patterns depended on the kind of ritual they were used in. This tradition is also of interest because of the mathematical ideas that are incorporated in it. Arithmetical relationships, progressions, symmetry, Euler graphs, and the (geometrical) determination of the greatest common divisor of two natural numbers are some of the mathematical ideas hidden in *sona* patterns.

Obviously, each of the design traditions mentioned above is incorporated into its respective culture and responds to its particular reality.

2 The tradition of Xysta

Xysta (singular *xysto* ; plural *xysta*) are a kind of graffiti that appears at the village of Pyrgi, one of the medieval villages of Chios. Although there are also a few houses in some other villages with *xysta*—mostly geometrical patterns constructed by traditional craftsmen on house façades—those in Pyrgi are considered a particular tradition (fig. 1).

The procedure for their construction is the following. First the craftsman plasters the façade of the house in one or two layers: the first makes the surface flat, while the second is the base for *xysta*.[1] While the material is fresh a layer of whitewash is added. The craftsmen subdivides the wet surface into zones, and in every zone appropriate patterns are designed. The pattern is then scratched with a fork into the whitewashed surfaces. The patterns that appear are the result of the contrast between the scratched whitewash and the plaster (fig. 2).

Fig. 1

Fig. 2

The main materials that are used for this procedure—depending on the time period—are different kinds of sand, mortar, whitewash and cement. The instruments that the traditional craftsmen use are only a lath, dividers with two points, and a fork. The lath serves two purposes: for the separation of the wall's surface in zones and for the construction of straight lines. The dividers are used for the construction of circular figures, while the fork is used for scratching some areas of the figures in a way the one area is dark (the scratched one) and the next white and so on.

As will be discussed below, this tradition is very important for the inhabitants' community and sense of identity. The fact that this is both a cultural practice as well as the application of interesting mathematical ideas in a traditional art form make *xysta* an interesting example of Ethnomathematics.

In this paper the following questions are discussed:

- How is the cultural context connected with this design tradition?
- What are the main mathematical ideas that we can see in these patterns?
- How is the construction of these patterns a result of informal cognition that craftsmen acquire through partnership?
- How could this be used for teaching some mathematical notions or practices?

3 Methodological issues: the method of the research

As Ethnomathematics lies in the confluence of mathematics and social (cultural) anthropology, the main methodology adopted comes from anthropology, namely ethnography. A commonplace of the researchers who explore cultural parameters is that "the place of emergence of cultural cognition is ethnography". It is argued that ethnographical research constitutes a particular characteristic of modern anthropology that differentiates it from the other social disciplines [Madianou 1999, 215].

A basic element in ethnography is research on site, with the main characteristic being participant observation. Participant observation combines participation in peoples' lives with a scientific distance that allows the precise observation and reporting of data. Also, participant observation is a kind of baptism in a culture.

In the framework of an ethnographic work, the researcher remains in the field as long as necessary in order to acquire access to aspects of life that could not otherwise be easily be approached in order to select data. In this type of research data can appear *a posteriori* as the result of meanings that are attributed in particular contexts and which researcher can see and interpret after he has been incorporated into the indigenous culture.

Participant observation is considered by some researchers as a method and by others as a research strategy or technique. Independently of the way we define participant observation, the majority of researchers consider it the most important as well as the most laborious method of anthropological research [Madianou 1999, 242]. The reason is that participation requires the involvement of the anthropologist in everyday activities and community life. Furthermore, communication through the local language is required. In fieldwork he has to observe and analyze the incidents in light of their everyday cultural relevance.

The procedure of interviewing is another important element. Open interviews are the most common type. Although they seem casual, because they have an implicit agenda—in comparison with the structured interview with its explicit agenda—these kind of interviews are useful for ethnographic research because they help the researcher understand the way people think and to compare the opinions of different people.

Another important aspect regarding fieldwork is entrance in the community. Since the ethnographer doesn't usually come from the community that he studies, how he or she approaches the members of the community is a significant issue.

The method that was selected for the present research was one with ethnographical characteristics. Although the time of my residence during the fieldwork was less than the usual, the tools of the research were very close to those of ethnography. Residence in the field, participant observation, and the informal interviews are some of the elements that determine the research presented here.

4 The fieldwork

Having visited Pyrgi a few times between 1993 and 2005, I had the sense that the geometrical patterns that the inhabitants create on the façades of their houses were of mathematical interest. The summer of 2005 I decided to stay in Pyrgi in order to study not only the kind of patterns but also the reasons why this tradition began and developed here. Thus most of the material in the present paper was the product of research on site.

4.1 The place and the people: past and present

Pyrgi, also known as the "painted village," is located in the north of Chios, one of the Aegean islands. Chios is well-known as the native land of the epic poet Homer. Today it is famous thanks to mastic, a product of the mastic tree. The inhabitants of Chios, especially those who come from the south, feel proud of their place because of the fact that while these trees also exist in other places around the world they don't produce mastic. Pyrgi is a mastic village.

Furthermore, Pyrgi is one of the medieval villages of Chios. What differentiates it from other medieval villages is the fact that it is substantially the same as it was six to seven centuries ago. Although there is an expansion of buildings, the main part of the settlement continues to be the same as it was in the past.

There is no sure information about the exact date of the settlement's construction. Among other writers, Konstantinos Sgouros [1937] asserts that the village existed before the possession of Genoa (1346-1566). Another historian, George Zolotas [1928], also believes that the main core of Pyrgi existed before the possession of Genoa. He also maintains that the inhabitants of Pyrgi and the nearby settlements were unified for safety purposes.

The architect-researcher Maria Xyda reports that the conquerors unified the settlements in order to fortify and organize the ex-Byzantine settlements that produced mastic into a single settlement [2000, 37]. Xyda estimates that the design of the village happened in another place. She notes that buildings such as churches were not included in the original design of the village, and thus claims that the design happened at Genoa. To support this argument she notes that the medieval villages of Chios were designed in the same way as Liguria's villages. The similarities between Chios's medieval villages and those of Liguria concern not only the urban layout but the constructive and morphological details of the houses as well as the use of similar stones [Xyda 2000, 38].

The German sojourner Hohann Michael Wansleben, who visited Chios in 1674, noted "Pyrgi is very well fortified and it has been built in the Italian way".

The shape of the settlement originally was a quadrangle. A small tower was built on each of the four apexes. The houses had neither windows nor doors on the external side, so they had a view only of the internal side of the settlement. The way those houses were built formed a wall around the settlement [Proiou 1992, 48]. Two main gates, in the north and the south, permitted access to the settlement. The houses were arranged like rings. At the boundaries of every ring the streets were made. These rings were linked by arcs [Xyda 2000, 41]. This form of the village was maintained until the beginning of the twentieth century, when it started to expand. As a result, today the boundaries are not distinguishable.

The houses of the old part of the village are very similar as far as the design and the material of construction are concerned. Usually the houses are constituted of three or four floors.

The type and the arrangement of the place dictate corresponding practises. First of all, because the inhabitants by definition live very close to each other, they have direct everyday contact with their neighbours, voluntary or not. What is of great importance for the present research is the fact that since the houses were narrow and dark, the inhabitants had to spend a lot of time outside. In other words, the whole social life of the population happens in the central square as well as in the streets around the square. The women meet their friends and their neighbours and do the household duties outside their homes, for example on the sidewalks in front of their houses, because there are no yards. At the time I was there I saw women preparing fresh beans, threading tomatoes—in order to dry them—and undertaking any other kind of activity they could do outside (figs 3-6).

Fig. 3-6. Social life in Pyrgi

The fact that they spent so much time outside their houses doing their everyday duties seems to have affected the way they realized the exteriors of them. The inhabitants initially constructed their houses with plain stones. As over time they improved their financial situation, they were able to plaster their houses in order to protect the walls from the weather. Later they started to add patterns to the facades of their houses for decoration purposes. So the practice of plastering, which was initially used simply for protection against the elements, developed into a way of decorating.

According to some sources, the patterns that they are used are derived from the carpets that Genovese people—the conquerors—used to put on the outsides of their houses for decorative purposes. After the Genovese left Pyrgi, the practice of decorating the façades with carpets was replaced by decorating façades with the patterns on plaster. Others consider the patterns of Capodocia in Turkey to be the inspiration (see [Xyda 2000]). The geographical location of Chios, and the Aegean islands in general, appears to explain the influences of both East and West.

4.2 The evolution of the style of *xysta*

As already mentioned, there is no sure information concerning either the date or the origin of the tradition of *xysta*. It is supposed that a constructive technique developed into a decorative one. When I asked Maria Xyda if there was an era at which the *xysta* were connected with a high status, she answered that the people who had the ability to plaster their houses and create *xysta*, in its earliest stages, were the wealthier people of the community.[2]

Although little is known about the date this tradition began, the techniques used from the beginnings up to the present is well known. Maria Xyda has classified the technique, the material, the patterns and the style, in general, into five categories [2000: 64-68].

- *Xysta* of the first period. At this time the patterns were only geometrical themes—limited types—and the plaster was made of river sand, lime and straw. The size of the patterns was similar to the size of the stones used as structural units (fig. 7).

Fig. 7

- *Xysta* with influences from the East. The *xysta* of this period were influenced from the Near East so there was a rich diversity in patterns. Another characteristic of this period was the tassels they had at the bottom, which represented carpets. Although the patterns are different from the first period the material used was the same (fig. 8).

Fig. 8

- *Xysta* of 1930-1940. The *xysta* at this time reflect all the previous influences. The material is not always the same since sea sand and cement were added (fig. 9).

| Fig. 9 | Fig. 10 |

- *Xysta* after the Second World War. The patterns become increasingly complicated, while at the same time they abandoned the use of colours. The material is the same except for the sand, which is now only from the sea, which is very close to the village (fig. 10).

- Contemporary *xysta*. The patterns are black and white. They aren't organized in units, but are more complicated and, in contrast with the previous periods, they extend over the entire surface of the facades. Furthermore the patterns are borrowed from some other traditions of craftsmanship, such as carpentry and blacksmithing. The patterns of this period cover the entire façade without taking into consideration the doors and windows. The materials that are now used are cement, a different type of sand that they can buy from the market, lime, and cinder (fig. 11).

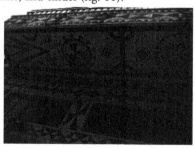

Fig. 11

4.3 The entrance in the community

After visiting Pyrgi a couple of times as a tourist, I decided that studying *xysta* could be of great interest in the context of the connections between culture and mathematics. This special tradition of *xysta* was important not only from the side of the construction and designing but also from the side of culture and mathematics. The majority of the patterns were geometrical constructions that were made by two simple instruments.

The first day I visited Pyrgi as a researcher, in the summer of 2005, I attempted my entrance in the community through a café located in the central square of the village in order to meet members of the community (*cafeneio* = Greek café). As mentioned above the social life of people takes place in the centre of the village. After this informal discussion with the inhabitants my informer Elias led me to observe some interesting patterns and also facilitated my contact with one of traditional craftsmen.

4.4 Material and data selection

The interviews with the inhabitants were informal and semi-structured. Through them I attempted mostly to understand elements about their identity as inhabitants of Pyrgi and the connection with the *xysta*. In contrast, the interviews with the craftsmen and the architects were more structured because more concrete answers concerning *xysta* were expected.

At Pyrgi, I had the chance to meet some very kind and helpful inhabitants who did their best to facilitate my research. Since the community is a small one, in a very short time, everybody had been informed that someone was interested in *xysta*. As a result, while I was walking down the streets or taking photos some inhabitants approached me and gave me any information about *xysta*.

In these discussions I heard several versions of the story of their origin as well as the date or period when this tradition started. Some of them consider the tradition to have come from the East (Turkey) and some others from the West (Italy). The location of Chios and, more generally, the area of Aegean Sea (indeed, the whole of Greece), allowed it to be influenced by both East and West.

Written documents selected from the local library supplemented the material of the research on site. In the library I found material concerning the tradition of *xysta* as well as the place and the people. Another very important resource for my research was my personal communication with the architect Maria Xyda. She comes from the island Chios and had conducted research about *xysta* in the framework of a European project. This resulted in her book, *The* Xysta *at Pygri of Chios*. In addition to personal communication, this book was of special interest for my research.

The ethnographical equipment used in the fieldwork were a camera to take some photos of the many designs of *xysta*, and paper and pencil in order to take notes during the fieldwork and to try some original analysis and thoughts. I used a tape recorder for the interviews.

4.5 Identity of Pyrgi's inhabitants and *xysta*

The majority of the inhabitants maintain that the main elements that distinguish their community from the 'others'—in the island and generally—are the *xysta* and their traditional dance, called *pyrgousiko*. Some added the traditional clothing, *pyrgoysiki*, as a distinguishing factor. The connection of the traditional clothing with the other two cultural peculiarities is noteworthy: according to their explanations, the designs on the sleeves of this clothing come from the *xysta*, although, as I noticed in the folklore museum, only a part of them had patterns similar to *xysta*. In any case, the fact that they speak this way, connecting these traditional elements and their identity, is itself of some importance.

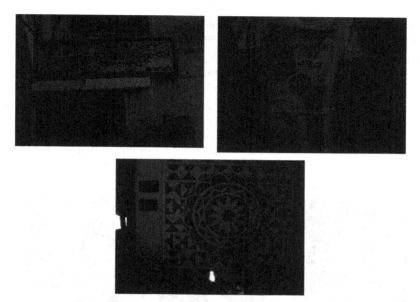

Figs. 12-14

The following discussion with an 80-year old man is characteristic of the importance of *xysta* for the members of community:

"Why do you like to have *xysta* at your house?"

"Because I'm *Pyrgouis* (=habitant of Pyrgi). Jesus Christ was born in the manger and the manger is what he remembers."

Many others answered the question about their interest in *xysta* in a similar way, saying that they like *xysta* because "these are our tradition". In some other cases it was tourist purposes that were emphasized: "The *xysta* is a means of promotion for Pyrgi, the place is famous because of its *xysta*".

The sign and the design on the t-shirt in figs. 12 and 13 are indicative of the connection of *xysta* with what is expected from tourism. Also, it was observed that some modern buildings, such as hotels, were decorated with *xysta*. In fig. 14 the interior of a modern hotel is shown. This hotel was located in the village closest to Pyrgi, which is its seaport. The majority of this village's habitants—including the owner of the hotel—came from Pyrgi. By using the *xysta* in an alternative way to decorate part of the hotel's interior he was declaring the continuity of the tradition.

5 Patterns and mathematical ideas

When studying the tradition of *xysta* in the framework of culture and mathematics, it is of great importance to understand how this tradition was incorporated and developed in this particular culture and what the meaning of it for the community is. On the other side, it is important to explore the mathematical ideas that are incorporated in them, noting that it is about informal mathematics, as *xysta* are products of craftsmen who have acquired this cognition through experience, without have been taught something in school.

A few indicative patterns presented here are going to be examined in order to help us to pick out interesting mathematical ideas. A main mathematical notion that is apparent in them is the construction of geometrical shapes such as rectilinear or circular figures.

Among others, the notion of transformation is one that appears very often in these patterns. The kind of transformation used in *xysta* patterns is isometry, since in all cases the shapes that are transformed do not change the distances between the points. The isometric objects are congruent; that is to say, we can turn one into the other just by sliding and flipping. The kinds of isometries that are found are translation, rotation, and reflection.

More analytically, in every picture we can observe the following mathematical ideas.

Figs. 15-16

The photos in figs. 15 and 16 show different parts of the façade of a church. Before starting the discussion about the geometrical figures it is interesting to notice some other elements on these pictures. For example, in the same place we can see religious symbols (such the cross) and other symbols (such as the half-moon and the point that refers to the East). Contradictions like this are characteristic of the Aegean islands, because they combine the culture of the West with that of the East, in particular the Greek Orthodox.

Observing the picture that covers larger surface we can see some zones that separate different motifs. The subdivision of the surface in parts could be considered as a set that is separated in sub-sets, a fact that also indicates a mathematical notion.

On the first clear zone we can see triangles that seem to be produced by a translation. Every triangle (black or white) is the union of two orthogonal triangles, while these orthogonal triangles come from the division of the rectangle in two equal parts by the diagonal. For this construction the informal notions that are implicitly used are the construction of the rectangle; the tracing of the diagonal; the fact that the diagonal divides the rectangle in two equal parts, and that alternating black and white triangles are congruent because they are derived from equal rectangles.

In the next zone the main notion is that of symmetry. In this there are two motives that are repeated. So, first of all we could speak about translation. Furthermore, in every motif there are two axes of symmetry: the one horizontal, the other vertical. Also the construction of circles (circular sectors), rectangles, and interstices between lines and circles are important mathematical ideas.

The other zones continue with similar mathematical notions and produce variants of the figures that have already been discussed above.

While the majority of the motives are constituted of rectilinear figures in some cases they are made of circular figures (fig. 17). As Maria Xyda noted, when the craftsmen didn't have enough space to develop a rectilinear figure, as for example in the space between a door and a window or under a balcony, they constructed circular motives [2000, 63]. The size of the circles, which they call "moons", depends on the available area. The only tool that is needed for the construction is dividers.

Fig. 17

In the preliminary stage for this pattern the craftsman constructed a quadrangle. After he had determined the centre of this, he traced the three concentric circles. Then, using a random point of the perimeter of the original circle as the centre, he traces a new circle whose radius is equal to that of the origin circle (all lines outside the original circle are later erased). He continued the procedure by tracing new circles, each time using the intersection of the previous circle with the original one as the new centre. After constructing the first six parts, which they call "daisy petals", by finding approximately the middle of one of the six arcs in which the circles have been divided, he continued with the same procedure. Scratching the lime of the common area of the petals as well as the area external them brings into evidence the final design. By continuing to scratch in the middle circle, the two rings are created. He finished the construction with the semi-circles on the outside of the circle.

In this pattern a lot of important implicit mathematical notions are present. First of all, there is the tracing of circles. For the construction of the main circle the centre of the quadrangle has to be determined. Behind the construction of equal arcs on the main circle is the equality between the meter of the arcs and the corresponding epicentre angle. Furthermore, in the main motif twelve axes of symmetry are noticed: six of them are diameters that connect two opposite points that are the intersection of the circles with the original circle and the other six are diameters than connect the middle of the arcs that are opposite.

An interesting façade is shown in fig. 18. It shows a *xysto* from the period 1930-1940 and it is reproduced from Maria's Xyda book [2000, 66]. As she notices, *xysta* of this period are the most naïf and characteristic since they combine the origins of the Pyrgi tradition and at the same time give solutions and perspectives for popular art.

Fig. 18

The patterns here are also a combination of geometrical figures. The constructions concern circles, semi-circles, quadrangles, equilateral and orthogonal triangles (half of a quadrangle), and the tracing of diagonals.

In the first clear zone, starting from the top the main figures are semi-circles. In this zone we can talk about translation as well as axial symmetry. In every motif two axes of symmetry are noticed: one horizontal and one vertical. Similarly in the next zone translation and axial symmetry are observed. The difference here is that there is only vertical axis in every motif. The equilateral and orthogonal triangles were produced by diagonals of the quadrangles.

In the next zone there is a more complicated design. The original figures are rectangles in which a horizontal line (parallel to the horizontal sides) and diagonals are traced. By scratching the triangles that are 1/4 or 1/8 of any rectangle we obtain these designs. By taking one rectangle as one unit, the next rectangle is produced by rotation. In the case that two adjacent rectangles are considered as one unit, we can talk about a translation.

It should be generally noted that all these patterns are constructed only with two tools: dividers and a straightedge without markings. Thus these constructions recall the only constructions that were acceptable in the mathematics of ancient Greece.

6 Some concluding notes

Xysta of Pyrgi is an interesting design activity because of both the significance for its inhabitants' culture and the mathematical ideas with which it is connected.

Current approaches of didactics of mathematics discuss the use of everyday mathematical cognition as well as examples of several cultures in teaching mathematical notions in the classroom. Patterns like these could be used in the introduction of mathematical notions such as transformation and symmetries.

By teaching mathematics through patterns, students can not only learn mathematics but can also understand that mathematics is a component of everyday life. Furthermore, they are motivated to find information about the corresponding community in which mathematical ideas are met in traditional activities, and thus see the connection between culture, cognition, and context.

Acknowledgments

The author thanks Chian Architect Maria Zyda for generous help during the fieldwork and to Professor Ubiratan D'Ambrosio for his remarks.

Notes

1. The number of layers of plaster on the façade depends on the technique as well as on the material that was used in the construction. The original *xysta* were only on stone houses, but now the majority of houses are brick.

2. Personal communication.

References

ASHER, M. and R. ASHER. 1981. *Code of the quipu*. Grand Rapids: University of Michigan Press.

BISHOP, A.J. 2002. *Mathematical Enculturation: A cultural Perspective on Mathematics Education*. Dordrecht, The Netherlands: Kluwer.

D'AMBROSIO, U. 2005. Preface. Pp. 10-17 in *Ethnomathematics: exploring the cultural dimension of mathematics and of mathematics education*, C. Stathopoulou. Athens: Atrapos.

GAY, J. and M. COLE. 1967. *The New Mathematics in an Old Culture*. New York: Holt, Rienhart and Winston.

GERDES, P. 1996. Ethnomathematics and Mathematics Education. Pp. 909-943 in *International Handbook of Mathematics Education*, J. Bishop et al. (eds). Dordrecht, The Netherlands: Kluwer Academic.

———. 1999. *Geometry from Africa: Mathematical and Educational Explorations*. Washington DC: The Mathematical Association of America.

———. 2005. *Geometrical aspects of Bora basketry in the Peruvian Amazon*. Mozambique: Mozambican Ethnomathematics Research Centre.

MADIANOU, D. 1999. *Culture and Ethnography: from the ethnographical realism to cultural criticism*. Athens: Greek Letters.

PINXTEN, R. 1983. Anthropology in the Mathematics Classroom? Pp 85-97 in *Cultural Perspectives on the Mathematics Classroom*, S. Lerman, ed. Dordrecht, The Netherlands: Kluwer Academic.

PROIOU, I. 1992. The history of Pyrgi. In *An Heirloom: Pyrgi of Chios*. Athens.

ZASLAVSKY, C. 1973. *Africa Counts*. Boston: Prindle, Weber and Schmidt.

SGOUROU, K. 1937. *History of Chios Island*. Athens.

ZOLOTAS, G. 1928. *History of Chios*. Athens.

ZASLAVSKY, C. 1994, "Africa Counts" And Ethnomathematics. *For the Learning of Mathematics* **14**, 2: 3-7.

XYDA, M. 2000. *The xysta of Pyrgi*. Chios: Alfa pi.

About the author

Charoula Stathopoulou teaches Mathematics and Didactics of Mathematics at the Special Education Department of University of Thessaly. Both Ethnomathematics and Ethnomathematics in conjunction with Mathematics Education are of special interest for her.

Rachel Fletcher

113 Division St.
Great Barrington, MA 01230
USA
rfletch@bcn.net

Keywords: Squaring the circle, descriptive geometry, Leonardo da Vinci, incommensurate values

Geometer's Angle

Squaring the Circle: Marriage of Heaven and Earth

Abstract. It is impossible to construct circles and squares of equal areas or perimeters precisely, for circles are measured by the incommensurable value *pi* (π) and squares by rational whole numbers. But from early times, geometers have attempted to reconcile these two orders of geometry. "Squaring the circle" can represent the union of opposing eternal and finite qualities, symbolizing the fusion of matter and spirit and the marriage of heaven and earth. In this column, we consider various methods for squaring the circle and related geometric constructions.

I Introduction

From the domed Pantheon of ancient Rome, if not before, architects have fashioned sacred dwellings after conceptions of the universe, utilizing circle and square geometries to depict spirit and matter united. Circular domes evoke the spherical cosmos and the descent of heavenly spirit to the material plane. Squares and cubes delineate the spatial directions of our physical world and portray the lifting up of material perfection to the divine.

Constructing these basic figures is elementary. The circle results when a cord is made to revolve around a post. The right angle of a square appears in a 3:4:5 triangle, easily made from a string of twelve equally spaced knots.[1] But "squaring the circle"—drawing circles and squares of equal areas or perimeters by means of a compass or rule—has eluded geometers from early times.[2] The problem cannot be solved with absolute precision, for circles are measured by the incommensurable value *pi* ($\pi = 3.1415927...$), which cannot be accurately expressed in finite whole numbers by which we measure squares.[3] At the symbolic level, however, the quest to obtain circles and squares of equal measure is equivalent to seeking the union of transcendent and finite qualities, or the marriage of heaven and earth. Various pursuits draw from the properties of music, geometry and even astronomical measures and distances. Each attempt offers new insight into the wonder of mathematical order. In this column, we consider methods for achieving circles and squares of equal perimeters, focusing on geometric approaches conducive to design applications and setting aside for now the problem of achieving circles and squares of equal areas.

Definitions:

The circle is the set of points in a plane that are equally distant from a fixed point in the plane.

The fixed point is called the center. The given distance is called the radius. The totality of points on the circle is called the circumference.

"Circle" is from the Latin *circulus*, which means "small ring" and is the diminutive of the Latin *circus* and the Greek *kuklos*, which mean "a round" or "a ring" [Liddell 1940, Simpson 1989].

The **circumference** is the line that forms the encompassing boundary of a circle or other rounded figure. The circumference (*c*) of a circle is $2\pi r$, where (*r*) is the length of the radius, or πd, where (*d*) is the length of the diameter. The area (*a*) of a circle is πr^2.

$$c = 2\pi r = \pi d$$
$$a = \pi r^2$$

The Latin for "circumference" is *circumferentia* (from *circum* "round, about" + *ferre* "to bear"), which is a late literal translation of the Greek *periphereia*, which means "the line around a circular body" or "periphery" [Liddell 1940, Simpson 1989].

The **square** is a closed plane figure of four equal sides and four 90° angles. "Square" is an adaptation of the Old French *esquare* (based on the Latin *ex-* "out, utterly" + *quadra* "square," which is from *quatuor* "four") [Harper 2001, Simpson 1989].

Perimeter is the term for the continuous line or lines that bound a closed geometrical figure, either curved or rectilinear, or of any area or surface. The perimeter (*p*) of a square is equal to four times the length of one of its sides (*s*):

$$p = 4s$$

The Latin for "perimeter" is *perimetros*, which means "circumference or perimeter," from the Greek *perimetros* (from *peri* "around" + *metron* "measure") [Lewis 1879, Simpson 1989].[*]

A circle of radius 1 is **equal in perimeter** to a square of side of $\pi/2$. (Each perimeter equals 2π.)

A circle of radius 1 is **equal in area** to a square of side $\sqrt{\pi}$. (Each area equals π.) (fig. 1)

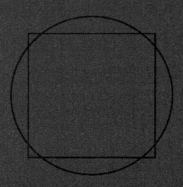
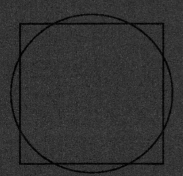

Circle of radius 1   Circle of radius 1
Square of side $\pi/2$ Square of side $\sqrt{\pi}$
Perimeters of circle and square = 2π Areas of circle and square = π

Fig. 1

II Vesica Piscis

A vesica piscis initiates our first technique for drawing a circle and square of equal perimeter, and is offered by John Michell.[5]

- Draw an indefinite horizontal line. Locate the approximate midpoint, at point O.

- Place the compass point at O. Draw arcs of equal radius that cross the horizontal line on the left and right, at points A and B.

- Set the compass at an opening that is slightly larger than before. Place the compass point at A. Draw an arc above and below, as shown.

- With the compass at the same opening, place the compass point at B. Draw an arc above and below, as shown.

- Locate points C and D where the two arcs intersect.

- Draw an indefinite vertical line through points C, O, and D.

Point O locates the intersection of the horizontal and vertical lines (fig. 2).

- Place the compass point at O. Draw a circle of indefinite radius, as shown.

- Locate point E where the circle intersects the vertical line, above.

- Place the compass point at E. Draw a circle of radius EO.

The horizontal line is perpendicular to the radius EO and tangent to its circle (fig. 3).

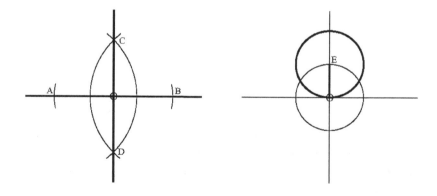

Fig. 2   Fig. 3

- Locate point E, then remove the circle whose center is point O.

- Locate point F where the remaining circle intersects the vertical line, as shown.

- Place the compass point at F. Draw a circle of radius FE.

- Extend the vertical line to the circumference of the circle (point G), as shown (fig. 4).

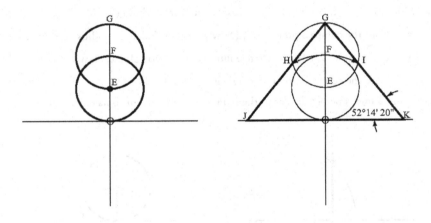

Fig. 4   Fig. 5

- Place the compass point at O. Draw an arc of radius OF that intersects the upper circle at points H and I, as shown.

- From point G, draw a line through point H that intersects the horizontal line (JK) at point J.

- From point G, draw a line through point I that intersects the horizontal line (JK) at point K.

- Connect points G, K and J.

The result is an isosceles triangle whose base angles measure 52°14'20" (52.2388...) (fig. 5).

- Place the compass point at O. Draw a circle of radius OJ.

- Place the compass point at J. Draw a half-circle of radius JO through the center of the circle (point O), as shown.

- Place the compass point at K. Draw a half-circle of radius KO through the center of the circle (point O), as shown.

- Locate points L and M, where the circle of radius OJ intersects the indefinite vertical line.

- Place the compass point at L. Draw a half-circle of radius LO through the center of the circle (point O), as shown.

- Place the compass point at M. Draw a half-circle of radius MO through the center of the circle (point O), as shown.

The four half-circles are of equal radius and intersect at points N, P, Q and R.

- Connect points N, P, Q and R.

The result is a square (fig. 6).

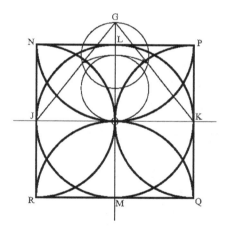
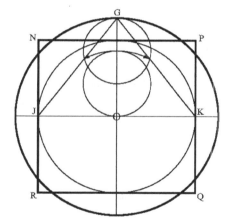

Fig. 6   Fig. 7

- Remove the four half-circles.

- Place the compass point at O. Draw a circle of radius OG.

If the radius OJ equals 1, then the radius (OG) of the large circle equals 1.29103....

If the value of π equals 3.14159, the circumference of the large circle equals 8.1117....

The side (NP) of the square equals 2 and the perimeter equals 8.0.

The circle and square are equal in perimeter within 1.4% (fig. 7).

III Double Vesica Piscis

Another method, based on a double vesica piscis, has been observed in traditional temple plans in India [Critchlow 1982, 30-31; Michell 1988, 40-42, 70-72].

- Repeat figure 2, as shown. Extend the horizontal and vertical lines in both directions.

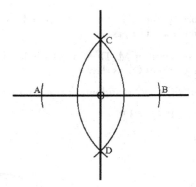

Fig. 2

- Place the compass point at O. Draw a circle of indefinite radius.

- Locate points E and F where the circle intersects the horizontal line.

- Locate points G and H where the circle intersects the vertical line (fig. 8).

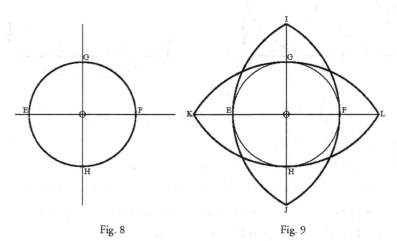

Fig. 8   Fig. 9

- Place the compass point at E. Draw an arc of radius EF that intersects the vertical line at points I and J.

- Place the compass point at F. Draw an arc of radius FE that intersects the vertical line at points I and J.

- Place the compass point at G. Draw an arc of radius GH that intersects the horizontal line at points K and L.

- Place the compass point at H. Draw an arc of radius HG that intersects the horizontal line at points K and L (fig. 9).

- Locate points M, N, P and Q where the four arcs intersect.

- Connect points M, N, P and Q.

The result is a square.

- Locate the circle of radius OE that is contained within the double vesica piscis.

If the radius (OE) of the circle equals 1 and the value of π equals 3.14159, the circumference of the circle equals 6.28318.

The side (MN) of the square equals 1.64575… and the perimeter equals 6.58300….

The circle and square are equal in perimeter within 4.8 % (fig. 10).

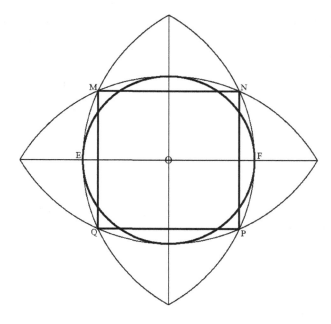

Fig. 10

IV Golden Section

This technique, offered by Robert Lawlor, utilizes the Golden Section, or Golden Ratio of 1 : *phi* or 1 : ϕ ($\phi = \sqrt{5}/2 + 1/2$), which translates numerically to the incommensurable ratio 1 : 1.618034….[6]

- Repeat figure 2, as shown.

- Place the compass point at O. Draw a circle of indefinite radius.

- Locate point E where the circle intersects the horizontal line, on the left.

- Place the compass point at E. Draw a circle of radius EO.

- Locate point F where the circle intersects the horizontal line, on the right.

- Place the compass point at F. Draw a circle of radius FO (fig. 11).

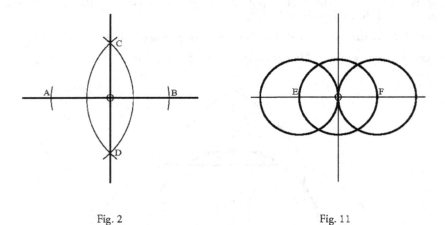

Fig. 2   Fig. 11

- Locate points G and H along the horizontal line, as shown.

- Place the compass point at O. Draw a circle of radius OH that encloses the three circles (fig. 12).

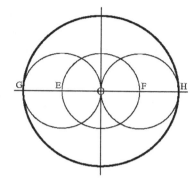

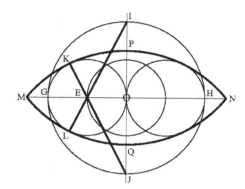

IP : PO :: PO : OH

$$\frac{1}{\Phi^2} : \frac{1}{\Phi} :: \frac{1}{\Phi} : 1$$

Fig. 12   Fig. 13

- Locate points I and J where the large circle intersects the vertical line.

- Draw a line from point I to point E. Extend the line IE to the circumference of the left circle (point L).

- Place the compass point at I. Draw an arc of radius IL, which intersects the extension of the horizontal diameter (GH) at points M and N.

- Draw a line from point J to point E. Extend the line JE to the circumference of the left circle (point K).

- Place the compass point at J. Draw an arc of radius JK, which intersects the extension of the horizontal diameter (GH) at points M and N.

If the radius (OH) of the large circle is 1, the radius (IL) of the arc (MN) equals *phi* (ϕ = $\sqrt{5}/2 + 1/2$ or 1.618034...), and half of the arc's long axis (ON) equals $\sqrt{\phi}$ (1.272019...).

If the short axis (PQ) of the arc equals 1, the diameters (GH and IJ) of the large circle equal ϕ^7 (fig. 13).

- Locate point I at the top of the vertical diameter (IJ) of the large circle.

- Place the compass point at I. Draw a half-circle of radius IO through the center of the circle (point O), as shown.

- Locate point H at the right end of the horizontal diameter (GH) of the large circle.

- Place the compass point at H. Draw a half-circle of radius HO through the center of the circle (point O), as shown.

- Locate point J at the bottom of the vertical diameter (IJ) of the large circle.

- Place the compass point at J. Draw a half-circle of radius JO through the center of the circle (point O), as shown.

- Locate point G at the left end of the horizontal diameter (GH) of the large circle.

- Place the compass point at G. Draw a half-circle of radius GO through the center of the circle (point O), as shown.

The four circles are of equal radius and intersect at points R, S, T and U.

- Connect points R, S, T and U.

The result is a square (fig. 14).

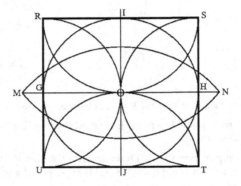

Fig. 14

- Remove the four half-circles.

- Place the compass point at O. Draw a circle of radius ON.

- Locate the radius (OH) of the smaller circle, as shown.

If the radius (OH) of the small circle equals 1, the radius (ON) of the large circle equals $\sqrt{\phi}$ (1.272019…).

If the radius (ON) of the large circle equals $\sqrt{\phi}$ and the value of π equals 3.14159, then the circumference of the large circle equals 7.99232….

The side (RS) of the square equals 2 and the perimeter equals 8.0.

The circle (of radius ON) and the square are equal in perimeter within 0.1% (fig. 15).[8]

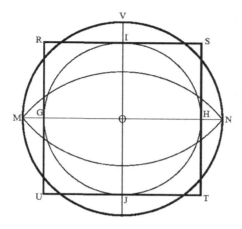

Fig. 15

V The Great Pyramid

- Connect points O, H, and V.

The result is a right triangle of sides 1 (OH) and √φ (OV). The hypotenuse (HV) is φ. Angle VHO equals 51°49'38" (51.827...) (fig. 16).

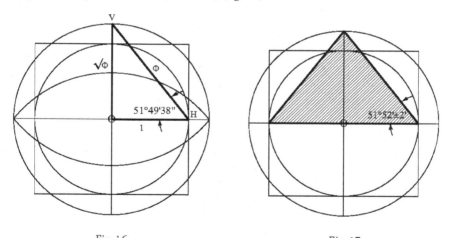

Fig. 16   Fig. 17

The Great Pyramid of Khufu, second king of the Fourth Dynasty (2613–2494 B.C.) and known to the Greeks as Cheops, is the largest of three pyramids at Gizeh, approximately eight miles from modern Cairo. The pyramids are built largely of limestone blocks with some granite, and erected near the edge of the limestone desert that borders the west side of the Nile valley. The approximate mean face angle of the Great Pyramid, based on calculations by Flinders Petrie, is 51°52'± 2' (51.866...) [Petrie 1990, xi, 12-13] (fig. 17).

Another method for achieving the proportions of the Great Pyramid, offered by John Michell, derives from a rhombus inscribed within a vesica piscis [1983, 158]. Fig. 18A presents an isosceles triangle whose base angle of 51°36'38" (51.61055...) approximates the face angle of the Great Pyramid.[9] Fig. 18B presents a square whose side equals the base of the triangle and a circle whose radius equals the height of the triangle. The circle and square are equal in perimeter within 0.8%.

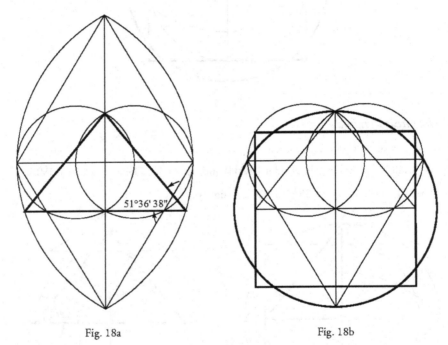

Fig. 18a   Fig. 18b

Kurt Mendelssohn proposes a practical method for achieving the Pyramid's proportions that utilizes the diameter and circumference, or revolution, of a rolling drum. The technique is not a true squaring of the circle, because it employs an instrument other than a compass and rule. But the relationship between circles and squares of equal perimeters is expressed in precise terms [Mendelssohn 1974, 73].

- Let the height (VO) of an isosceles triangle (VHG) equal the length of four drums stacked tangent to one another.

- Let half the base (OH) of the triangle equal the length of the circumference, or one revolution of one drum.

- Place the compass point at O. Draw a circle of radius OV.

- Place the compass point at O. Draw a circle of radius OH.

- Draw a square (RSTU) about the circle of radius OH.

If the diameter of each drum equals 1, the radius (OV) of the large circle equals 4, and the radius (OH) of the smaller circle equals π.

The circumference of the large circle (radius OV) and the perimeter of the square (RSTU) each equal 8π precisely.

If π equals 3.14159, angle VHO equals 51°51'14" (51.854...), which is approximately the mean face angle of the Great Pyramid (fig. 19).

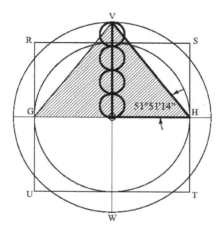

Fig. 19

VI Relative Measures of Earth and Moon

One solution for squaring the circle, offered by John Michell, derives from actual astronomical measures. The construction is based on a circle of radius 5040, representing in miles the combined mean radii of the circles of the earth (3960) and the moon (1080).[10]

- Draw a circle representing the earth (mean radius 3960 miles) and a circle representing the moon (mean radius 1080 miles) tangent to one another, as shown (fig. 20).

- Draw a square about the circle of radius 3960 (earth).

- Draw a circle about the combined radii of 3960 (earth) and 1080 (moon), or 5040.

If π equals the Archimedean value of 22/7, the circle of radius 5040 and the square drawn about the "earth" circle of radius 3960 are exact (31,680) (fig. 21).

The measures in Michell's construction express added meaning when converted to different scales and units of measure, suggesting that the different measuring systems are interrelated. For example, 31,680 in miles is both the circumference of the circle drawn on the combined radii of the earth and moon and the perimeter the square containing the circle of the earth alone. But the number 31,680 in furlongs is the mean radius of the earth (3960 miles) and in inches is half a mile [Michell 1988, 33, 173].

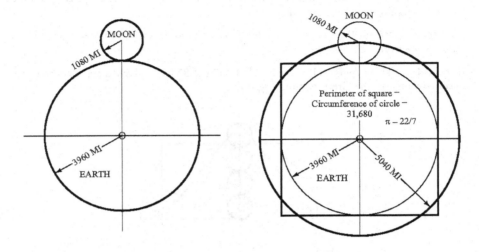

Fig. 20   Fig. 21

In miles, the radius of the moon is 1080. One hundred and eight (108) is the atomic weight of silver, the metal we traditionally associate with the moon, whose silvery surface reflects the light of other bodies. The mean distance between the earth and sun is four times 10,800 diameters of the moon. The Roman half-pace of 1.216512 feet divides 108,000,000 times into earth's mean circumference. The Hebrew calendar divides the hour into 1080 units, called *chalaki*, based on the number of breaths one is presumed to take in one hour. The number 108 appears in religious symbolism, such as the 108 beads in the Hindu or Buddhist rosary, 10,800 stanzas in the *Rigveda*, and 10,800 bricks in the Indian fire altar [Michell 1988, 180-181].[11]

The ideal city-state Magnesia, envisioned by Plato in the *Laws*, consists of 5040 individual allotments of land to be distributed among 5040 citizens [Plato 1961: *Laws* V, 737e, 1323]. In miles, 5040 is the combined radii of the earth and moon. The number 5040 is the product of 1 x 2 x 3 x 4 x 5 x 6 x 7 and contains sixty individual divisions. The number 7920, which is the mean diameter of the earth in miles, is the product of 8 x 9 x 10 x 11. Thus, the product of 5040 and 7920 is the product of the numbers 1 through 11 [Michell 1988, 109-110].

Michell observes this arrangement of astronomical measures in temple plans throughout history. St John's New Jerusalem, the celestial city described at the beginning of the Christian Era in the New Testament book of *Revelation*, is based on a square of 12 x 12 furlongs, containing a circle of circumference 14,400 cubits. This translates to a circle of 7920 feet in diameter and 24,883.2 feet in circumference, compared to the earth's diameter of 7920 miles and circumference of 24,883.2 miles. The perimeter of a square circumscribing the circle is 31,680 feet [Michell 1988, 24-25].

The New Jerusalem plan is an idealized vision of heaven and earth, but it can also be a house, temple, village, city, or entire world-order. Glastonbury, England, where some believe Druidic mysteries yielded to Christianity in the west, is associated with Arthurian legend. The original width of St. Mary's Chapel, built at Glastonbury Abbey, is 39.6 feet. Its footprint, a 1 x √3 rectangle, would be circumscribed by a circle of diameter 79.2 feet. The perimeter of the circumscribing square would be 316.8 feet [Michell 1988, 28-29].

Stonehenge, the megalithic monument in Salisbury, England, also reproduces the dimensions of St. John's city on a reduced scale of 1:100, when expressed in feet. Thus, the mean circumference of the outer circle of sarsen stones is 316.8 feet. The diameter of the inner ring of bluestones is 79.2 feet [Michell 1988, 31, 173].

Michell's geometric symbol contains additional layers of meaning, which may be accessed through "gematria," a term from medieval Kabbalah adopted from the Greek *geômetria* or "geometry" that associates the letters of Greek, Hebrew and other ancient alphabets with numerical values, musical tones and vibrations, colors, and geometric images [Liddell 1940, Simpson 1989]. In this way, numbers and measures convey musical, astronomical and mythological content. For example, by gematria, the Greek το αγιον πνευμα or *to hagion pneuma*, which means "the Earth Spirit," and το γαιον πνευμα or *to gaiôn pneuma,* which means "the Holy Spirit," each yield the number 1080.[12]

VII The Heptagon

Definitions:

The **regular polygon** is a plane figure in which all sides are equal and all interior angles are equal.

In a regular polygon with (*n*) sides, the **interior angle** is (180-360/*n*) degrees. The sum of the polygon's interior angles is (180*n*-360) degrees.

"Polygon" is via late Latin from the Greek *polugônos* (from *polu* "many" + *gônia* "corner, angle") and *polugonos*, which means "producing many at a birth, prolific" [Harper 2001, Liddell 1889, Liddell 1940].

"Heptagon" is from the Greek *heptagônos* (from *hepta* "seven" + *gônia* "corner, angle" [Harper 2001, Liddell 1940, Simpson 1989]. A regular heptagon contains seven equal sides that meet at seven equal interior angles of 128°34'17" (128.57142...).

A regular heptagon cannot be constructed precisely with a compass and rule, but one approximate construction relates to the squaring the circle. Let us begin with the method for squaring the circle that is based on the Golden Section.

- Repeat figure 16, as shown.
- Locate points G and H where the small circle intersects the horizontal diameter.
- Locate points V and W where the large circle intersects the vertical diameter.
- Connect points V, H, W and G.

The result is a rhombus (fig. 22).

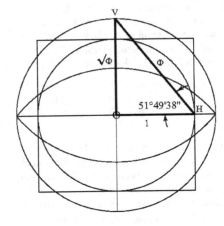

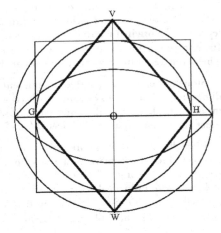

Fig. 16

Fig. 22

- Place the compass point at G. Draw a circle of radius GO.

- Place the compass point at H. Draw a circle of radius HO (fig. 23).

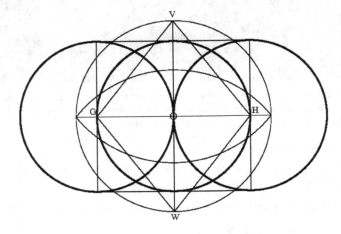

Fig. 23

- Locate point A where the line VH intersects the right circle, as shown.

- Locate point B where the line WH intersects the right circle, as shown.

- Connect points OA and OB (fig. 24).

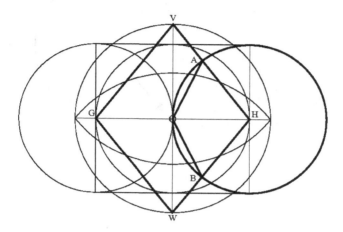

Fig. 24

Lines OA and OB locate two approximate sides of a regular heptagon inscribed within the right circle.

- Place the compass point at A. Draw an arc of radius AO that intersects the right circle at point C, as shown.

- Place the compass point at B. Draw an arc of radius BO that intersects the right circle at point D, as shown.

- Place the compass point at C. Draw an arc of radius CA that intersects the right circle at point E, as shown.

- Place the compass point at D. Draw an arc of radius DB that intersects the right circle at point F, as shown.

- Connect points A, C, E, F, D, B and O.

The result is a heptagon that approximates a precise regular heptagon (fig. 25).[13]

- Repeat the process within the left circle, as shown.

Angle AOB equals 128°10′22″ (128.17277…). The interior angles of a true regular heptagon equal 128°34′17″ (128.57142…).

The constructed heptagon approximates a true heptagon within 0.3% (fig. 26).

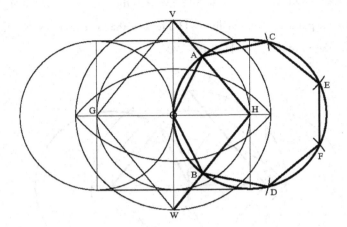

Fig. 25

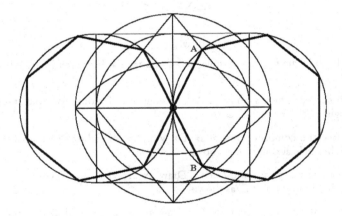

Fig. 26

VIII Brunés Sacred Cut

In a previous column, we examined the "sacred cut," so named by Tons Brunés for its ability to generate a circle and square of nearly equal perimeters and to divide the side of a square into seven nearly equal parts. The square grid contains a center square, four smaller corner squares, and four 1 : √2 rectangles (fig. 27).[14]

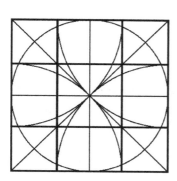

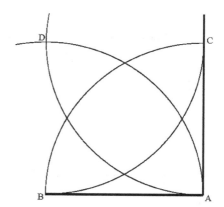

Fig. 27   Fig. 28

Brunés' technique for squaring the circle is based on the observation that a quarter-arc drawn on half the diagonal of a square and the diagonal of half the square are equal in length within 0.6% [Brunés 1967: I, 73-74, 93-94; Watts 1987, 268-269; Watts 1992, 309-310].

- Draw a horizontal baseline AB equal in length to one unit.

- From point A, draw an indefinite line perpendicular to line AB that is slightly longer in length.

- Place the compass point at A. Draw a quarter-arc of radius AB that intersects the line AB at point B and the open-ended line at point C.

- Place the compass point at B. Draw a quarter-arc (or one slightly longer) of the same radius, as shown.

- Place the compass point at C. Draw a quarter-arc (or one slightly longer) of the same radius, as shown.

- Locate point D, where the two quarter-arcs taken from points B and C intersect.

- Place the compass point at D. Draw a quarter-arc of the same radius that intersects the line AC at point C and the line AB at point B (fig. 28).

- Connect points A, B, D and C.

The result is a square (ABDC) of side 1.

- Locate points E and F where the quarter-arcs intersect above and below, as shown.

- Draw the line EF.

- Extend the line EF in both directions to points G and H on the square.

Line GH divides the square (ABDC) in half.

- Locate points I and J where the quarter-arcs intersect on the left and right, as shown.

- Draw the line IJ.

- Locate the point O where the lines GH and IJ intersect.

Point O marks the midpoint of the line GH and the center of the square (fig. 29).

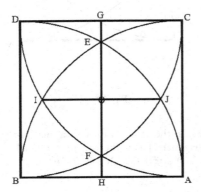

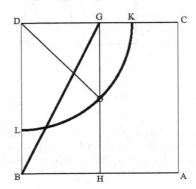

Diagonal GB = √5/2 or 1.11803...
Quarter-arc KL equals (π√2)/4 = 1.11072...

Fig. 29   Fig. 30

- Remove the four quarter-arcs.

- Mark the location of point O and remove the line IJ.

- Draw the semi-diagonal GB.

- Locate points D and O.

- Draw the line DO.

- Place the compass point at D. Draw a quarter-arc of radius DO that intersects the line DC at point K and the line DB at point L.

If the side (AB) of the square is 1, the diagonal (GB) equals √5/2, or 1.11803... .

The radius DO equals 1/√2 or √2/2.[15]

If π equals 3.14159, the quarter-arc (KL) drawn on radius DO equals (π√2)/4 or 1.11072....

The diagonal (GB) and the quarter-arc (KL) are equal in length within 0.6% (fig. 30).

Diagonal GB = √5/2 or 1.11803...

Quarter-arc KL equals $(\pi\sqrt{2})/4 = 1.11072...$

Figure 31 presents a square (ABDC) of side AB, a square of side GB, and a circle of radius OD.

If the side (AB) of the square ABDC is 1, the perimeter of the square of side GB equals 4.47213... and the circumference of the circle of radius OD equals 4.44288....

The circumference of the circle and the perimeter of the square (of side GB) are equal in length within 0.6% (fig. 31).

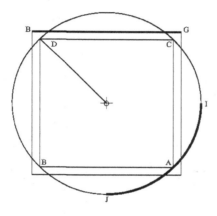

Fig. 31

IX Leonardo's Vitruvian Man

In 1490, Leonardo da Vinci produced the illustration we know today as "Vitruvian Man." The study depicts the set of ideal human proportions proposed by Vitruvius in *De architectura* (*Ten Books on Architecture*), in which the adult male figure is proportioned to a circle and a square. Neither Vitruvius nor Leonardo propose a circle and a square of equal measure, but in Leonardo's interpretation the two are superimposed, to dramatic effect.[16]

The Vitruvian canon of human proportion is well known:

> The center and midpoint of the human body is, naturally, the navel. For if a person is imagined lying back with outstretched arms and feet within a circle whose center is at the navel, the fingers and toes will trace the circumference of this circle as they move about. But to whatever extent a circular scheme may be present in the body, a square design may also be discerned there. For if we measure from the soles of the feet to the crown of the head, and this measurement is compared with that of the outstretched hands, one discovers that this breadth equals the height, just as in areas which have been squared off by use of the set square [Vitruvius 1999: III, i, 47].

Following Vitruvius, Leonardo locates the navel of the human figure at the center of a circle whose circumference bounds the figure's outstretched arms and legs. The figure's total height and arm span are equal in length and measure the edge lengths of a square. The center of the square locates the genitals. Quarter divisions locate the nipples, the base of the knees, the junction of forearm and upper arm, and the width of the shoulders. An eighth division locates the bottom of the chin. When the arms are raised in line with the top of the head, the middle fingers indicate where the circle and square intersect (fig 32).[17]

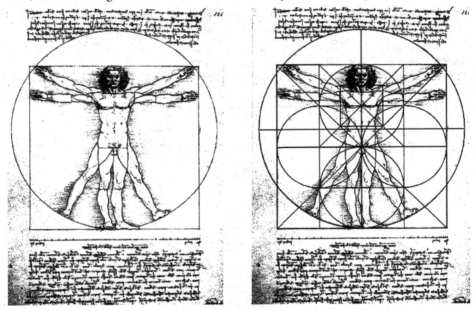

Fig.32. Image: Canon of proportions from Vitruvius's *De architectura* (*Ten Books on Architecture*). Leonardo da Vinci, c. 1490. Venice: Accademia. Geometric overlay: Rachel Fletcher.

The intended result is a harmony of individual parts and the whole:

> And so, if Nature has composed the human body so that in its proportions the separate individual elements answer to the total form, then the ancients seem to have had reason to decide that bringing their creations to full completion likewise required a correspondence between the measure of individual elements and the appearance of the work as a whole [Vitruvius 1999: III, i, 47].

The Vitruvian scheme may be applied to habitats and dwellings of every kind—from houses, temples and cities to the cosmos itself—and is one of several attempts throughout history to express human proportions in precise geometric terms.[18] Whether the results are exact, such efforts reflect a basic human need to perceive a coherent, harmonious world. If geometry is the art of reconciling diverse spatial elements, the quest to square the circle is an art of the highest order.

Notes

1. [See Fletcher 2005b, 44]. Thales of Miletus is said to have learned this technique in Egypt [Padovan 1999, 60-61].
2. J. J. O'Connor and E. F. Roberson provide a concise history, beginning with the Egyptian Rhind papyrus that was scribed by Ahmes and was based on an original dating from 1850 B.C. or earlier. A square nearly equal in area to that of a circle is accomplished when the square is constructed on 8/9 of the circle's diameter [O'Connor and Roberson 1999; van der Waerden 1983, 170-172].

   To E. W. Hobson and others, "squaring the circle" is the name for circles and squares of equal areas and is also known as a circle's quadrature. Hobson's term for circles and squares of equal perimeters is "rectifying the circle." [Hobson 1913, 4-5]. Occasionally, the *ad quadratum* construction, in which a circle is inscribed within a square, or a square within a circle, is named "squaring the circle," even though this does not produce figures of equal measure. [See Fletcher 2005b, 45-49.]
3. In some ways, the quest to square the circle parallels the history of π, which Hobson traces through three historical periods. The first "geometric period" from prehistory through the sixteenth century consisted of producing approximate values for π from geometric constructions. One method attributed to Socrates' contemporary Antiphon begins with a square inscribed in a circle. The sides of the square are bisected, then those of the octagon that results, then the 16agon and so on, until the polygon is indistinguishable from a circle. Socrates' contemporary Bryson improved on this method by considering circumscribed and inscribed polygons together. Another method derives from a theorem by Archimedes (287-212 B.C.), which states that the area of a circle equals the area of a right-triangle whose short side equals the radius of the circle and whose long side equals the circumference [Hobson 1913, 10-11, 15-19; Smith 1958, 302-307].

   Hobson's second "analytical" period, from the mid-seventeenth century, applied analytical processes, specifically trigonometric functions, to solve the problem of π. Not until the third period, from the mid-eighteenth until the late nineteenth century, was π shown to be truly irrational or transcendental [Hobson 1913, 12-13, 43-57].
4. See [Fletcher 2004, 95-96] for more on the circle and [Fletcher 2005b, 35-37] for more on the square.
5. [1969, xxxi]. See [Fletcher 2004, 96] for more on the vesica piscis.
6. [1982, 74-76.] See [Fletcher 2006, 67] for more on the Golden Ratio.
7. See [Fletcher 2005a, 151-153], to derive a regular pentagon from this construction.
8. Note this construction is based on fact that $\sqrt{\phi}$ and $4/\pi$ are nearly exact.

$$\sqrt{\phi} = 1.272019\ldots$$
$$4/\pi = 1.273239\ldots$$

   Put another way,

$$\pi = 3.14159\ldots.$$
$$4/\sqrt{\phi} = 3.144605\ldots$$

9. To construct the isosceles triangle, draw a vesica piscis from two circles, as shown, such that the center of one circle coincides with the circumference of the other. Next, draw a larger vesica piscis whose short axis equals the width of the two circles. Inscribe a rhombus within the larger vesica piscis. The height of the isosceles triangle equals the long axis of the smaller vesica piscis. The base of the isosceles triangle passes through two points where the rhombus and the small circles intersect.
10. Modern estimates for astronomical distances vary, but their averages are nearly identical to Michell's figures, whose calculations utilize two approximate values of π: the Archimedean value of 22/7 (= 3.14285...) and Fibonacci's approximation of 864/275 (= 3.14181...) [Beckmann, 1971, 66,84]. Michell's measure for the mean radius of the earth is based on a value of π equal to 864/275. For a full account see [Michell 1988, 100-106].

11. Ancient Greek and Hebrew numbering systems do not recognize zero or "0." Therefore, the numbers 108, 1080 and 10,800, although different in quantity, share the same qualitative value [Bond 1977, 6].

12. [Michell 1988, 181.] The individual letters in ΤΟ ΑΓΙΟΝ ΠΝΕΥΜΑ (το αγιον πνευμα), "the Holy Spirit" are: [(300)Τ + (70)Ο] + [(1)Α + (3)Γ + (10)Ι + (70)Ο + (50)Ν] + [(80)Π + (50)Ν + (5)Ε + (400)Υ + (40)Μ + (1)Α]. These add to (370 + 134 + 576) and distill to 1080. The individual letters in ΤΟ ΓΑΙΟΝ ΠΝΕΥΜΑ (το γαιον πνευμα), "the Earth Spirit" are [(300)Τ + (70)Ο] + [(3)Γ + (1)Α + (10)Ι + (70)Ο + (50)Ν] + [(80)Π + (50)Ν + (5)Ε + (400)Υ + (40)Μ + (1)Α]. These add to (370 + 134 + 576) and distill to 1080 [Bond 1977, 6].

13. Side EF is slightly shorter than the others.

14. To derive fig. 26, see [Fletcher 2005b, 56-61].

15. The calculation of the diagonal GB is based on the Pythagorean theorem, such that $BH^2 + HG^2 = GB^2 [(1/2)^2 + 1^2 = (5/4)^2]$. Thus, GB = $\sqrt{5}/2$. The calculation of DO is based on the fact that the diagonal of a square of side 1 is equal to $\sqrt{2}$. See [Fletcher 2005b, 44-45] for more on the Pythagorean theorem.

16. Lionel March reconciles the circle and square in the Vitruvian figure through a regular octagon whose base equals the base of the square, and whose half-chord equals the diameter of the circle. The margin of error is approximately 0.57%. If the base of the square and the base of the octagon share the exact location, the center of the octagon and the circumference of the circle nearly coincide. Robert Lawlor proposes a less precise interpretation based on the Golden Section [Lawlor 1982, 59; March 1998, 106-108].

17. Leonardo notes these and other alignments in *The Theory of the Art of Painting*. In addition, he says, "If you open your legs so much as to decrease your height 1/14 and spread and raise your arms till [*sic*] your middle fingers touch the level of the top of your head you must know that the center of the outspread limbs will be in the navel and the space between the legs will be an equilateral triangle" [Richter 1970, I, 182]. March observes the equilateral triangle in an analysis of his own [March 1998, 107].

   For this analysis, the source image of Vitruvian Man was manipulated to correct for distortion in aspect ratio, possibly the result of irregular paper shrinkage. The manipulated image presents a true circle.

18. See [Fletcher 2006, 83-84] for more on human proportions.

References

BECKMANN, Petr. *A History of π (PI)*. New York: St. Martin's Press.

BRUNÉS, Tons. 1967. *The Secrets of Ancient Geometry – and Its Use*. 2 vols. Copenhagen: Rhodos.

BOND, Frederick Bligh and Thomas Simcock Lea. 1977. *Gematria: A Preliminary Investigation of the Cabala*. London: Research Into Lost Knowledge Organization.

CRITCHLOW, KEITH. 1982. *Time Stands Still: New Light on Megalithic Science*. New York: St. Martin's Press.

FLETCHER, Rachel. 2004. Musings on the Vesica Piscis. *Nexus Network Journal* 6, 2 (Autumn 2004): 95-110.

———. 2005a. SIX + ONE. *Nexus Network Journal* 7, 1 (Spring 2005): 141-160.

———. 2005b. The Square. *Nexus Network Journal* 7, 2 (Autumn 2005): 35-70.

———. 2006. The Golden Section. *Nexus Network Journal* 8, 1 (Spring 2006): 67-89.

HARPER, Douglas, ed. 2001. *Online Etymological Dictionary*. http://www.etymonline.com/

HOBSON, E. W. 1913. *"Squaring the Circle": A History of the Problem*. Cambridge: Cambridge University Press. Reprint. Michigan Historical Reprint Series. University of Michigan University Library. No date.

LAWLOR, Robert. 1982. *Sacred Geometry: Philosophy and Practice*. New York: Thames and Hudson.

LEWIS, Charlton T. and Charles Short, eds. 1879. *A Latin Dictionary*. Oxford: Clarendon Press. Perseus Digital Library Project. Gregory R. Crane, ed. Medford, MA: Tufts University. 2005. http://www.perseus.tufts.edu

LIDDELL, Henry George and Robert Scott, eds.1889. *An Intermediate Greek-English Lexicon*. Oxford. Clarendon Press. Perseus Digital Library Project. Gregory R. Crane, ed. Medford, MA: Tufts University. 2005. http://www.perseus.tufts.edu

LIDDELL, Henry George and Robert Scott, eds. 1940. *A Greek-English Lexicon*. Henry Stuart Jones, rev. Oxford: Clarendon Press. Perseus Digital Library Project. Gregory R. Crane, ed. Medford, MA: Tufts University. 2005. http://www.perseus.tufts.edu

MARCH, Lionel. 1998. *Architectonics of Humanism: Essays on Number in Architecture*. London: Academy Editions.

MENDELSSOHN, Kurt. *The Riddle of the Pyramids*. New York: Praeger.

MICHELL, John. 1969. *The View Over Atlantis*. New York: Ballantine.

———. 1983. *The New View Over Atlantis*. London: Thames and Hudson.

———. 1988. *The Dimensions of Paradise: The Proportions and Symbolic Numbers of Ancient Cosmology*. San Francisco: Harper and Row.

O'CONNOR, J. J. and E. F. Robertson. 1999. Squaring the Circle. School of Mathematics and Statistics, University of St. Andrews: St. Andrews, Fife, Scotland. http://www-history.mcs.st-andrews.ac.uk/HistTopics/Squaring_the_circle.html

PADOVAN, Richard. 1999. *Proportion: Science, Philosophy, Architecture*. London: E & FN Spon.

PETRIE, W. M. Flinders. 1990. *The Pyramids and Temples of Gizeh*. 1885. Reprint. London: Histories & Mysteries of Man Ltd.

PLATO 1961. *The Collected Dialogues of Plato Including the Letters*. Edith Hamilton and Huntington Cairns, eds. Princeton: Bollingen Series LXXI of Princeton University Press.

RICHTER, Jean Paul, ed., 1970 *The Notebooks Works of Leonardo da Vinci*. 2 vols. 1883. Reprint. New York: Dover Publications.

SIMPSON, John and Edmund Weiner, eds. 1989. *The Oxford English Dictionary*. 2nd ed. OED Online. Oxford: Oxford University Press. 2004. http://www.oed.com/

SMITH, David Eugene. 1958. *History of Mathematics*. Vol. II. 1925. Reprint. New York: Dover Publications.

VAN DER WAERDEN, B. L. 1983. *Geometry and Algebra in Ancient Civilizations*. Berlin: Springer-Verlag.

VITRUVIUS. 1999. *Ten Books on Architecture*. Trans. Ingrid D. Rowland, Ed. Ingrid D. Rowland and Thomas Noble Howe. Cambridge: Cambridge University Press.

WATTS, Carol Martin and Donald J. Watts. 1987. Geometrical Orderings of the Garden Houses at Ostia. *Journal of the Society of Architectural Historians* **46**, 3 (1987): 265-276.

WATTS, Carol Martin and Donald J. Watts. 1992. The Role of Monuments in the Geometrical Ordering of the Roman Master Plan of Gerasa. *Journal of the Society of Architectural Historians* **51**, 3 (1992): 306-314.

About the geometer

Rachel Fletcher is a theatre designer and geometer living in Massachusetts, with degrees from Hofstra University, SUNY Albany and Humboldt State University. She is the creator/curator of two museum exhibits on geometry, "Infinite Measure" and "Design By Nature". She is the co-curator of the exhibit "Harmony by Design: The Golden Mean" and author of its exhibition catalog. In conjunction with these exhibits, which have traveled to Chicago, Washington, and New York, she teaches geometry and proportion to design practitioners. She is an adjunct professor at the New York School of Interior Design. Her essays have appeared in numerous books and journals, including "Design Spirit", "Parabola", and "The Power of Place". She is the founding director of Housatonic River Walk in Great Barrington, Massachusetts, and is currently directing the creation of an African American Heritage Trail in the Upper Housatonic Valley of Connecticut and Massachusetts.

B. Lynn Bodner

Mathematics Department
Monmouth University
West Long Branch
New Jersey 07764 USA
bodner@monmouth.edu

Keywords: Mathematics, computer science, science, art, architecture, sculpture, music, dance, theatre, education

Conference report

Bridges 2006: Mathematical Connections in Art, Music, and Science

4-9 August 2006, London

Abstract. B. Lynn Bodner reports on the Bridges 2006 conference.

Since 1998, practicing mathematicians, artists, musicians and scientists have been coming together at the annual Bridges Conference to share ideas and enthusiasm for a commonly held interest in the mathematical connections existing among the fields of mathematics, computer science, science, art, architecture, sculpture, music, dance, theater and education. This year's conference was hosted by the London Knowledge Lab, an interdisciplinary research lab (http://lkl.ac.uk), and the Institute of Education, a postgraduate institution, both affiliated with the University of London (in the United Kingdom). The six-day conference (from August 4 through August 9, 2006) included invited plenary speaker presentations in the mornings, parallel contributed paper sessions and workshops in the afternoons, a visual art exhibit open throughout the entire day, a musical evening, family day and various excursions. This review will briefly discuss each of these and highlight some of the author's most memorable experiences.

The plenary speakers, Jacqui Carey, ("Bridging the Gap – a Search for a Braid Language"), Xavier De Kestelier and Brady Peters ("The Work of Foster and Partners Specialist Modeling Group"), Michael Field ("Illuminating Chaos – Art on Average"), Louis Kauffman ("The Borromean Rings – A Tripartite Topological Relationship"), Peter Randall-Page ("Collaborating on the Integration of Sculpture and Architecture in the Eden Project"), Carlo Sequin ("Patterns on the Genus-3 Klein Quartic"), Caroline Series ("Non-Euclidean Symmetry and Indra's Pearls"), and Simon Thomas ("Love, Understanding and Soap Bubbles") all gave extraordinarily interesting presentations during the morning sessions of the conference. Picking one as a highlight is extremely difficult, however, having said that, this author found the presentation of Xavier De Kestelier and Brady Peters, members of the Specialist Modeling Group at Fosters and Partners Architects, especially intriguing since some of the structural designs they discussed, including the Swiss Ré Gerkin (fig. 1) and the Greater London Authority building, visibly add to the distinctive sky line of central London. The web links of individual plenary speakers may be found at http://www.lkl.ac.uk/bridges/programme.html.

Fig. 1. The Swiss Ré Gerkin (in the background)

Each afternoon, four parallel contributed paper sessions and 'Bridges for Teachers, Teachers for Bridges' workshops on a wide range of topics were held concurrently, making for difficult decisions on the part of conference goers as to which event to attend. There were two sessions of seven contributed papers each on the connections between mathematics and music, a session of seven papers on Islamic art, and nine other sessions (consisting of sixty-three papers in total) on other disparate mathematical art topics. The eighty-minute workshops involved various topics for teachers. With so much from which to choose, the author (as was the case for all conference participants) was only able to sample one fifth of the available sessions, and so it is impossible to discuss the highlights of these sessions. Instead, one is encouraged to read the short abstracts which may be found at http://www.lkl.ac.uk/bridges/abstracts.html and the conference *Proceedings,* containing the full text of the refereed papers, published by and available from Tarquin Publications (www.tarquinbooks.com).

Another major and very popular feature of this year's conference was the Bridges Art Exhibit, displaying the largest collection of mathematical art (over 150 pieces from 54 contributors) since Bridges' inception. Robert Fathauer, the Art Exhibit Coordinator, and Anne Burns, who created and maintains the extensive website, also served as jurors of the artwork, along with Nat Friedman, Reza Sarhangi and John Sharp. The collection, which includes sculptures, prints, and quilts, is well worth a look at:

http://myweb.cwpost.liu.edu/aburns/bridges06/bridges06.html.

On Monday August 7, the Bridges Musical Evening, which was a free event open to the public (thanks in part to the support of Sibelius Software), combined musical performances and short lectures "aiming to illustrate – through music – how mathematics is intertwined with human activity and creativity." (For more information on most of these presentations, see the following webpage: http://www.lkl.ac.uk/bridges/musical.html). The most rousing moment of the evening occurred when the audience was invited to join Paco Gomez and Godfried Toussaint in a performance of Steve Reich's "Clapping Music." The audience, in concert with Paco, clapped a constant rhythm with their hands while Godfried clapped shifts in the rhythmic pattern by one unit of time after a fixed number of repetitions, until eventually all were clapping in unison again. For those of us musically challenged, this was much harder to do for the entire length of the piece than this description sounds!

Fig. 2. Constructing George Hart's paper model during Family Day

The Bridges Family Day on Wednesday August 9, an event for "children of all ages from 5 to 95," was organized in conjunction with the Royal Institute of Great Britain (www.rigb.org) and planned as "a day to inspire, engage and motivate; to show that Maths really can be 'fun', especially when art is involved too" (fig. 2). Two parallel mathematics masterclasses (on perspective, anamorphic art, juggling, and Celtic and African art) and a Zometool workshop were held in the morning and a series of mathematics "hands on" art activities were available in the afternoon. (See http://www.lkl.ac.uk/bridges/familyday.html for a description of the program and some of the spontaneous mathematical art activities planned.) To cite just a few, Jacqui Carey, one of the plenary speakers of the conference and a braidmaking specialist, had us creating our own beautiful braids in no time; George Hart and Bradford Hansen-Smith led us in the construction of paper models and sculptures, and David Mitchell had us folding amazing three-dimensional shapes and the regular polyhedra.

As an adjunct to the conference, participants were given a wide range of excursion choices for Saturday August 5, including an exclusive tour of mathematical sites at

Cambridge University, conducted by the distinguished mathematician Michael Longuet-Higgins; a Bloomsbury walking tour led by David Singmaster, mathematician and metagrobologist; a visit to the Hindu temple, Shri Swaminarayan Mandir, led by Phillip Kent, research officer in Mathematics Education of the London Knowledge Lab; an Islamic Kensington tour led by John Sharp, mathematics writer, educator and visiting fellow of the London Knowledge Lab; a walking tour of mathematical and tourist sights in central London, led by Patricia Wackrill, and a walk from Westminster to Trafalgar Square, led by Penelope Woolfitt. Another optional bus excursion on Tuesday August 8 involved visits to the New Art Centre Sculpture Park and Gallery, Roche Court, which contains (among other things) *Warp and Weft* (fig. 3), a wonderful granite glacial boulder sculpture by Peter Randall-Page, one of the plenary speakers; Salisbury Cathedral, one of England's great medieval buildings, which contains (among other sights) carved stone polyhedra on the Tomb of Sir Thomas Gorges; and Stonehenge, a key historic site in Britain. For more information and web links on the various excursions, please see http://www.lkl.ac.uk/bridges/excursions.html.

Fig. 3. *Warp and Weft* by Peter Randall-Page at the New Art Centre
Sculpture Park and Gallery, Roche Court

All in all, this year's Bridges Conference was one of the best ever, with over 200 participants from all over the world, 150 works of art on display at the Art Exhibit, presentations and workshops on a wide variety of topics, and engaging mathematical art activities for all. For information on previous Bridges conferences and also next year's conference to be held in San Sebastian, Spain, please visit the Bridges home page at http://www.sckans.edu/~bridges/.

Acknowledgment

A Spanish version of this conference report was published in *Matematicalia*, October 2006.

About the reviewer

B. Lynn Bodner is an associate professor of mathematics at Monmouth University in New Jersey, USA, having taught a wide variety of undergraduate mathematics courses for twenty-three years. She especially enjoys teaching classes on the geometries, the mathematics of artistic design, and the historical development of mathematics. Her most recent scholarship interest involves the study of medieval geometric Islamic art which incorporates ideas from all three of these areas. Her webpage may be found at: http://mathserv.monmouth.edu/coursenotes/bodner/bodner.htm.

Sylvie Duvernoy

Via Benozzo Gozzoli, 26
50124 Florence ITALY
sduvernoy@kimwilliamsbooks.com

Keywords: Guarino Guarini,
Chapel of the Holy Shroud,
Baroque architecture, projective
geometry, mechanics

Symposium report

Guarino Guarini's Chapel of the Holy Shroud in Turin: Open Questions, Possible Solutions

18-19 September 2007, Turin, Italy

Abstract. Sylvie Duvernoy reports on the symposium on Guarino Guarini and the Chapel of the Holy Shroud, held in September 2006 in Turin.

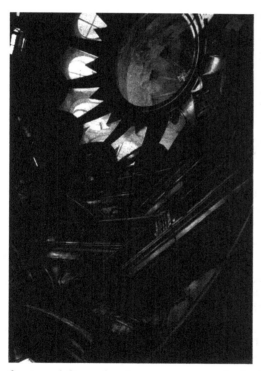

In mid-September this year, the Archivio di Stato of Turin hosted an international symposium dedicated to the study of the Chapel of the Holy Shroud in Turin and its designer, Guarino Guarini, organized by Kim Williams and Franco Pastrone, and sponsored by the Archivio di Stato and the Direzione per i beni culturali e paesaggistici del Piemonte.

The Chapel of the Holy Shroud is an astonishing construction in which architectural design, decoration and static requirements are united in complex relationships that are not easy to understand and clarify. It is a major monument of Italian Baroque architecture and its architect – Guarino Guarini – is among the great figures of the Italian *Seicento*, together with Bernini and Borromini.

Because some pieces of the interior marble cornice had fallen, the Chapel was closed to the public in the early 1990s, and inquiries into requirements for its stability and maintenance were made. Analyses showed that the cause of the fall of the marble pieces was not due to structural problems, but instead to the intrinsic weakness of the Frabosa marble, whose veins and mineral structure, over time, lead to cracking. Only slight repairs and a thorough cleaning were therefore necessary. Unfortunately, on the night of 11 April 1997, when the required maintenance was already complete and the scaffolding that had been erected inside the chapel during the work was ready to be dismantled, a fire broke out in the scaffolding itself, and developed over the course of two hours before being discovered – thus devastating the entire building – before the fire brigade reached the monument. This world-famous catastrophe resulted in enormous damage to the Chapel:

the whole stone covering of the interior was ruined, having exploded due to the change of temperature from the heat of the fire and the cold of the water used to extinguish it. The glass of the windows burst out, and some of the structural iron cables broke. Were it not for the prompt action taken by the firemen themselves (the only people allowed to work in no-security conditions) the dome would have totally collapsed.

The events that occurred that night, together with the results of the early studies that were undertaken in the months that followed, were already reported in the *Nexus Network Journal* (vol. 6 no.2, 2004) in the transcription of an interview with Mirella Macera (Superintendent for architectural, landscapes, and historical monuments of Piedmont), Fernando Delmastro and Paolo Napoli, (the architect and the engineer in charge of the preliminary studies for the restoration project) conducted by Kim Williams.

The September symposium was intended to be a sort of report on the work and studies in progress, now that the judiciary process is over and that the operational phase of the restoration can start: a kind of pause for reflection during which scholars from many countries gathered in order to share information and ideas. The talks and discussions followed two main themes: the Chapel itself (its structural analysis and architectural design), and the Chapel's designer: Guarino Guarini.

As the first speaker, Mirella Macera described to the audience the updated situation of the so-called *cantiere della conoscenza*, i.e., the latest progress in knowledge that has been made while cleaning and classifying every single stone piece of the interior veneer of the Chapel.

Just after this introduction, Paolo Napoli brilliantly explained the complex structural system of the whole building, pointing out the differences that exist between the original drawings by Guarini and the final form of the dome itself. In these discrepancies lies the very core of the restoration problem, since the orientation of the restoration project will depend on their interpretation. The historical documentation is fragmentary and discontinuous. Guarini's drawings do not fully describe the construction process, and no written data regarding the modifications and adjustments that he had to do while completing the structure is available. Debate about his original intentions and options is therefore open. And since the general cultural attitude in Italy regarding architectural restoration is strongly linked to the faithful adhesion to the original intention of the designer, the importance of this discussion and the related conclusion must not be underestimated. Which of the two has to be restored: the final static situation or the initial project, even if it is less efficiently resistant?

The talks by Santiago Huerta and Elwin Robison also focused on the static aspects of the Chapel, although from a more theoretical and general point of view. Huerta pointed out some interesting analogies between the structure and Gothic architectural principles.

Although the Chapel of the Holy Shroud is Guarino Guarini's *chef d'œuvre*, it is not his only architectural work. Ugo Quarello presented research on the church of San Lorenzo in Turin, especially focusing on the restoration work undertaken during the eighteenth century. Pietro Totaro spoke about the façade of the church of the Santissima Annunziata dei Teatini in Sicily and its influence on the Sicilian Baroque architecture.

In addition to being an architect, Guarini also was a theologian and a religious (a Theatine), a mathematician and a philosopher. His literary works are particularly impressive. His treatises include works on philosophy (*Placita philosophica*, 1665); mathematics (*Euclides adauctus et methodicus mathematicaeque universalis*, 1671); architecture (*Modo di misurare le fabriche*, 1674; *Trattato di fortificazione che hora usa in Fiandra, Francia et Italia*, 1676, and *Disegni d'architettura civile ed ecclesiastica*, 1686); cosmology (*Compendio della sfera celeste*, 1675; *Leges temporum et planetarum*, 1678; *Coelestis mathematicae*, 1683). Some of the scientific treatises were presented and discussed during the symposium: *Coelestis mathematicae* (presented by Patricia Radelet-de Grave) and *Euclides adauctus et methodicus mathematicaeque* (presented by Clara Silvia Roero and Anastasia Cavagna and Michele Maoret).

But Guarini's major essay related to architecture remains the *Disegni d'architettura civile ed ecclesiastica*, a foremost treatise in the specialized international literature of the seventeenth century, illustrated by Joël Sakarovitch, who pointed out how Guarini mastered the techniques of projective geometry, and his skills as a draftsman. The nexus between drawing and design (between "project" and "projection") was addressed by Michele Sbacchi. The delicate question of the reciprocal influence between design and representation is a recurrent discussion topic in Nexus conferences and workshops. Asserting that the plan drawings by Guarini are clear orthogonal projections of the spatial geometry above, Sbacchi adds a further contribution to this debate. The participants at the symposium were able to see some of the original drawings by Guarini and other contemporary architects, which belong to Turin's Archivio di Stato.

James McQuillan and Vasileios Ntovros each presented an interpretation of the symbolic aspects of the architecture and its geometrical pattern, but while McQuillan preferred to link his research to the historical and cultural context of the Baroque period and to Guarini's own writings, Ntovros based his own investigation on the concept of "folding and unfolding" inspired from Gilles Deleuze's definition in the book *FOLD, Leibniz and the Baroque* (University of Minnesota Press, 1992). No contradiction arose from the results of the two analyses, showing how scientific research may be supported either by traditional methodologies or by modern approaches to produce convincing evidence. The rigour of the research alone guarantees its scientific value.

The two-day Turin symposium was not supposed to have any influence whatsoever on the future orientation of the restoration project of the Chapel of the Holy Shroud, but the quality of the works that were presented – and of the discussions that followed – surely contributed some valuable information to the *cantiere della conoscenza* about the Chapel and its designer.

The symposium was made possible by contributions from the Associazione Subalpina Mathesis, Torino, Comune di Vigliano Biellese, the Department of Mathematics of the University of Turin, the Assessorato alla Cultura della Regione Piemonte, and Kim Williams Books. Publication of the *Proceedings* is planned for 2007.

About the reviewer

Sylvie Duvernoy is the Book Review Editor of the *Nexus Network Journal*.

Kay Bea Jones

Knowlton School of
Architecture
189 Brown Hall
190 W 17th Ave.
Columbus OH 43210 USA
jones.76@osu.edu

Keywords: Modern
architecture, Franco Albini,
Renzo Piano, Zero Gravity

Exhibit Review

Zero Gravity. Franco Albini. Costruire le Modernità

Milan Triennale
28 September-26 December 2006

Abstract. Kay Bea Jones reviews the exhibit of the work of Franco Albini in Milan.

On a spectacular, warm Friday evening in early fall, throngs of the fashionably clad pushed into the great Rationalist hall of the Milan Triennale to hear Andrea Cancellato, Triennale director, curator Fulvio Irace, architect Renzo Piano, and media personality Vittorio Sgarbi inaugurate the centennial exhibition of the prolific and astylistic work of architect Franco Albini (1905-1977). Both Irace and Piano, who installed the show with Franco Origoni, dedicated the event to the recently deceased Milanese designer, Vico Magistretti. Magistretti has left his own mark on Milan's modern culture in the form of mass produced, sleek, smart, everyday furniture. I asked Vico only a few years ago what he thought of the design work of our evening's protagonist: "Ahhh, Franco Albini," he sighed, "He was born too soon."

Piano and Origoni's installation design is light, frugal, and almost invisibile – appropriate to set off the work of the home town architect who consistently sparked poetic expression from pragmatic circumstances. Piano's address and catalog statement acknowledge that Albini has been his source of design rigor, material ethos, and tectonic sophistication throughout his career. Piano, after dropping out of the Florence school of architecture after his third year, stalked Albini until he took him into the studio as an apprentice from 1960-63. As a young draftsman in the office where apparently no one spoke, Piano drew detail upon detail for projects that included the Rinascente Department Store in Rome and the Palazzo Rosso Gallery renovation in Genoa. Albini's design method and built objects are renowned for their rigorous craft, and Piano "stole daily with his eyes wide open." The contemporary Italian architect, who like his predecessor became world

1590-5896/07/010155-4 DOI 10.1007/s00004-006-0036-4
© Kim Williams Books, Turin

renowned for his museums, came to light for the Centre Georges Pompidou in Paris and has produced the Menil and Cy Twombly Museums in Houston; the High Museum expansion in Atlanta; The Beyeler Foundation in Basil, Switzerland; the Morgan library in Manhattan; and is currently working on the Chicago Art Institute and the conflict-ridden Whitney. Piano concluded by saying that he was glad to celebrate "Albini's poetic which for me has been so formative."

If the sublime modern San Lorenzo Museum (1954) that holds the Treasures of the Duomo of Genoa were more accessible, it might be as important to the history of modern architecture as Mies van der Rohe's Crown Hall or Kahn's Trenton Bath House. Philip Johnson surely knew of it when he designed his underground painting gallery at his New Canaan estate (1965). The buried treasure held in four cellular rooms banded in local grey stone (*promontorio*) with cast concrete spines is presented in "Zero Gravity" via a large original model, meticulous drawings that render each stone coursing, and period construction shots along with new photographs that honor the 1995 renovation. The success of this Triennale exhibition depends largely on the wealth of material provided by the Albini studio and archives. The range of the architect's accomplishments is made explicit through museums, installations, interiors, housing, furniture, and urban proposals, and the detailed presentation of this complex assemblage verifies what Manfredo Tafuri called Albini's "technically faultless vocabulary." No other architect so thoroughly and consciously designed each and every room as though it were the essential element of modern architecture. In fact, Albini's many domestic interiors for clients and exhibits during the 1930s were Albini's means of research. They eventually comprise a thesis that defines his later work: that modernity begins not with the object building, but with the perfect room. Walls become abstract planes from which paintings are removed and suspended on steel armatures. Removing heavy picture frames allowed images to float in the open spaces of galleries or dwellings. Custom furniture made of glass yields transparent views and reflected surfaces. Gravity is dynamically challenged as the room's contents float and stairs hover above the ground. Albini's formal language that so gracefully manipulates the constraints of construction to render poetic each element's component parts relies on concept rather than modern style for its unique architectural expression. The architect's son, Marco, notes that these carefully developed elements are critical to understanding the significance of his father's lessons for objects that take on a life of their own. For example, the versatile and elegant composite column of the Veliero bookshelf became the interior column system for the Brera installation, the Palazzo Rosso gallery, and the Olivetti showroom. Through repetition and evolution Albini modelled the accretion of space showing that a modern place is the sum of well-conceived and crafted parts to generate a greater cohesive whole.

"Zero Gravity" explains the key motif in the Triennale assemblage of Albini's oeuvre. Even the notable talking heads testify from suspended plasma screens. Among them Renzo Piano, Marco Albini, Vittorio Gregotti, Albini collaborators Enea Manfredini, Matilde Baffa and others discuss the role Albini played in forming a modern ethos. The contents of the exhibit are divided into eight sections: Bachelor Machines; The New City: Milan and Rational Architecture; Atmospheric Spaces: The Architecture of Exhibitions; Dwelling Objects; Rooms of Memory; Modernity and Tradition; The Museums between Albini and Scarpa; and the Technology of the City, each curated by a different scholar or pair of scholars. In spite of the subdivision of interests gleaned from disassociated scholarship, the show holds together. "Zero Gravity" is the first of three simultaneous exhibits titled

"Costruire le Modernità'" (Constructing Modernity). The other two shows to be hosted in Genoa and Turin later this year will feature the coincident careers of Ignazio Gardella and Carlo Mollino, respectively, each also born in 1905.

"Zero Gravity" is distinguished by the quality of original material in the show – Albini's transparent radio (1938), the original columns of the Veliero glass bookshelf (1940), static only when loaded with books, Cassina's reconstruction of the same artifact, the table for the aviator Ferrarin (1932), several chairs with their cherry-red upholstery, and countless original drawings in ink or graphite on delicate yellow tracing paper that show the refined clarity of Albini's design method. High-quality original black and white photographs enlarged and suspended to fill the space are the staple representation for built work, especially for ephemeral designs no longer accessible. One is left to wonder why the original Triennale installations from 1936 (Room for a Man) and 1940 (Living room for a Villa) were not reconstructed for this occasion, especially since the same venue is hosting the current homage. In the interest of situating Albini across the modern century, comparisons are forged between Albini and other Italian modernists, including Persico, Mollino, Scarpa, Michelucci, Libera, and Moretti. The Milanese are clearly enthralled with their home-grown architect, and recognize his many contributions to the built and social fabric of the city, especially noteworthy popular housing, the Villetta Pestarini, the MSA (Il Movimento di Studi per l'Architettura) debates after World War II, collaborations for a better city in the proposals for Milano Verde and the A.R. (Architetti Reuniti) plans, and the award-winning lines 1 and 2 of the Milan subway. Yet a missed opportunity of the DARC-co-sponsored Triennale tribute lies in the insular nature of the endeavor – only Italian critics were summoned and only Italian architects were sought for comparison. Had Albini's accomplishments relative to those of Lou Kahn, Philip Johnson, and Lina Bo Bardi (working in Brazil) been examined with fresh perspectives by scholars outside the Milanese circle, the show would likely reach a wider audience, including some who are not yet aware of Albini's role in constructing modernity. As so often happens with anniversary shows, there is too much worship and too little critical interrogation. Albini collaborated throughout his long career, and while various collaborators are mentioned, none are considered unduly influential in his constantly developing vocabulary. The most noticeable oversight is the lack of consideration of Franco Helg, who worked with Albini from 1951 until his death and carried on the studio until 1989.

The exhibition catalog, *Zero Gravity Franco Albini Costruire le Modernità*, published by The Milan Triennale and Mondadori Electa S.p.A., was edited by Federico Bucci and Fulvio Irace (2006).

About the reviewer

Kay Bea Jones is an associate professor at the Ohio State Austin E. Knowlton School of Architecture where she has taught for twenty-one years. She began the KSA abroad studies program in Italy and teaches a traveling seminar that considers cross-cultural uses of public space. She has written and lectured widely about travel pedagogy and Italian modern and contemporary architecture. Jones recently published *Publi-city: techniques for the survival of public space* (with Pippo Ciorra, and Beatrice Bruscoli), a discussion of their intercultural pedagogical collaborations. Her architectural research on alternatives to market rate housing has resulted in the recent Buckeye Village Community Center at Ohio State, which she designed with George Acock and Andrew Rosenthal. The building recently received the national 2006 EDRA Places Design Award and an American Institute of Architects 2006 Merit Award. Jones has published and lectured widely on modern Italian architecture. She has collaborated with colleagues at the Milan Polytechnic to bring the traveling exhibition of the museums and installations of Franco Albini to eight North American venues, where she has lectured about his complex modernity regarding room as the unit element of modern architecture. Her forthcoming book on this is titled *Suspending Modernity: The Architecture of Franco Albini*.

Printed in the United States
By Bookmasters